Higher

Biology

D1390147

2004 Exam

2005 Exam

2006 Exam

2007 Exam

2008 Exam

Leckie✕Leckie

© Scottish Qualifications Authority
All rights reserved. Copying prohibited. No part of this publication may be reproduced, stored in a retrieval system, or transmitted in any form or by any means, electronic, mechanical, photocopying, recording or otherwise.

First exam published in 2004.
Published by Leckie & Leckie Ltd, 3rd Floor, 4 Queen Street, Edinburgh EH2 1JE
tel: 0131 220 6831 fax: 0131 225 9987 enquiries@leckieandleckie.co.uk www.leckieandleckie.co.uk

ISBN 978-1-84372-671-5

A CIP Catalogue record for this book is available from the British Library.

Leckie & Leckie is a division of Huveaux plc.

Leckie & Leckie is grateful to the copyright holders, as credited at the back of the book, for permission to use their material.
Every effort has been made to trace the copyright holders and to obtain their permission for the use of copyright material.
Leckie & Leckie will gladly receive information enabling them to rectify any error or omission in subsequent editions.

[BLANK PAGE]

FOR OFFICIAL USE

Total for
Sections
B and C

X007/301

NATIONAL
QUALIFICATIONS
2004

WEDNESDAY, 19 MAY
1.00 PM – 3.30 PM

BIOLOGY
HIGHER

Fill in these boxes and read what is printed below.

Full name of centre

Town

Forename(s)

Surname

Date of birth
Day Month Year

Scottish candidate number

Number of seat

SECTION A—Questions 1–30 (30 marks)

Instructions for completion of Section A are given on page two.

SECTIONS B AND C (100 marks)

1   (a)   All questions should be attempted.

(b)   It should be noted that in **Section C** questions 1 and 2 each contain a choice.

2   The questions may be answered in any order but all answers are to be written in the spaces provided in this answer book, and must be written clearly and legibly in ink.

3   Additional space for answers and rough work will be found at the end of the book. If further space is required, supplementary sheets may be obtained from the invigilator and should be inserted inside the **front** cover of this book.

4   The numbers of questions must be clearly inserted with any answers written in the additional space.

5   Rough work, if any should be necessary, should be written in this book and then scored through when the fair copy has been written.

6   Before leaving the examination room you must give this book to the invigilator. If you do not, you may lose all the marks for this paper.

SCOTTISH
QUALIFICATIONS
AUTHORITY

SECTION A

Read carefully

1   Check that the answer sheet provided is for Biology Higher (Section A).

2   Fill in the details required on the answer sheet.

3   In this section a question is answered by indicating the choice A, B, C or D by a stroke made in **ink** in the appropriate place in the answer sheet—see the sample question below.

4   For each question there is only **one** correct answer.

5   Rough working, if required, should be done only on this question paper—or on the rough working sheet provided—**not** on the answer sheet.

6   At the end of the examination the answer sheet for Section A **must** be placed inside the front cover of this answer book.

Sample Question

The apparatus used to determine the energy stored in a foodstuff is a

A   respirometer

B   calorimeter

C   klinostat

D   gas burette.

The correct answer is **B**—calorimeter. A **heavy** vertical line should be drawn joining the two dots in the appropriate box in the column headed **B** as shown in the example on the answer sheet.

If, after you have recorded your answer, you decide that you have made an error and wish to make a change, you should cancel the original answer and put a vertical stroke in the box you now consider to be correct. Thus, if you want to change an answer D to an answer B, your answer sheet would look like this:

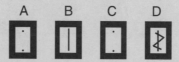

If you want to change back to an answer which has already been scored out, you should enter a tick (✓) to the **right** of the box of your choice, thus:

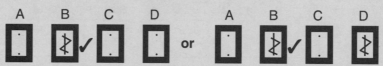

SECTION A

All questions in this section should be attempted.

Answers should be given on the separate answer sheet provided.

1. Which of the following processes requires infolding of the cell membrane?

   A Diffusion

   B Phagocytosis

   C Active transport

   D Osmosis

2. The diagram shows the fate of sunlight landing on a leaf.

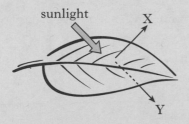

   Which line in the table below identifies correctly the fate of sunlight represented by X and Y?

	X	Y
A	transmission	reflection
B	absorption	transmission
C	reflection	transmission
D	reflection	absorption

3. Which of the following colours of light are mainly absorbed by chlorophyll a?

   A Orange and violet

   B Blue and red

   C Blue and green

   D Green and orange

4. The graph shows the effect of temperature on the rate of reactions in the light dependent stage in photosynthesis.

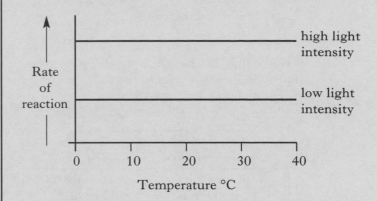

   From the graph, it may be deduced that

   A enzymes are not involved in controlling these reactions

   B enzymes act most effectively at high intensities of light

   C at the high intensity of light, carbon dioxide is the limiting factor

   D the rate of the reaction increases with increase in temperature.

5. The graph shows the effect of increasing light intensity on the rate of photosynthesis.

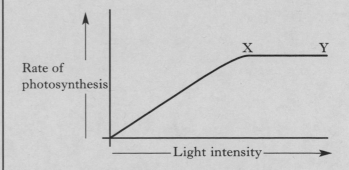

   Two environmental factors which could limit the rate of photosynthesis between points X and Y are

   A light intensity and oxygen concentration

   B temperature and light intensity

   C temperature and carbon dioxide concentration

   D carbon dioxide concentration and light intensity.

6. In respiration, the sequence of reactions resulting in the conversion of glucose to pyruvic acid is called

A the Krebs cycle

B the citric acid cycle

C glycolysis

D the cytochrome chain.

7. Which line in the table describes correctly both aerobic respiration and anaerobic respiration in human muscle tissue?

	Aerobic Respiration	*Anaerobic Respiration*
A	There is a net gain of ATP	Carbon dioxide is not produced
B	There is a net gain of ATP	Oxygen is required
C	Carbon dioxide is produced	There is a net loss of ATP
D	Lactic acid is formed	Ethanol is formed

8. Cyanogenesis in *Trifolium repens* is a defence mechanism against

A water loss

B fungal infection

C bacterial invasion

D grazing.

9. A sex-linked gene carried on the X-chromosome of a man will be transmitted to

A 50% of his male children

B 50% of his female children

C 100% of his male children

D 100% of his female children.

10. The inheritance of eye colour in *Drosophila* is sex-linked and the allele for red eyes (R) is dominant to the allele for white eyes (r).

The progeny of a cross were all red-eyed females and white-eyed males.

What were the genotypes of their parents?

A X^rX^r X^RY

B X^RX^r X^RY

C X^RX^r X^rY

D X^RX^R X^rY

11. Black coat colour in cocker spaniels is determined by a dominant gene (B) and red coat colour by its recessive allele (b). Uniform coat colour is determined by a dominant gene (F) and spotted coat colour by its recessive allele (f).

A male with a uniform black coat was mated to a female with a uniform red coat. A litter of six pups was produced, two of which had uniform black coat colour, two had uniform red coat colour, one had spotted black coat colour and one had spotted red coat colour.

The genotypes of the parents were

A BBFf × bbFf

B BbFf × bbFF

C BbFf × BbFf

D BbFf × bbFf.

12. A tall plant with purple petals was crossed with a dwarf plant with white petals. The F_1 generation were all tall plants with purple petals.

The F_1 generation was self pollinated and produced 1600 plants.

Which line in the table identifies correctly the most likely phenotypic ratio in the F_2 generation?

	Tall purple	Tall white	Dwarf purple	Dwarf white
A	870	325	305	100
B	870	0	0	730
C	400	400	400	400
D	530	260	270	540

13. The table below shows the percentage recombination frequencies for four genes present on the same chromosome.

Gene pair	% recombination frequency
P and Q	33
R and Q	40
R and S	32
P and R	7
Q and S	8

Which of the following represents the correct order of genes on the chromosome?

A	Q	P	S	R
B	P	Q	S	R
C	Q	S	P	R
D	P	Q	R	S

14. Which of the following describes the term non-disjunction?

A The failure of chromosomes to separate at meiosis.

B The independent assortment of chromosomes at meiosis.

C The exchange of genetic information at chiasmata.

D An error in the replication of DNA before cell division.

15. Which of the following is true of polyploid plants?

A They have reduced vigour and the diploid chromosome number.

B They have increased vigour and the diploid chromosome number.

C They have reduced vigour and sets of chromosomes greater than the diploid chromosome number.

D They have increased vigour and sets of chromosomes greater than the diploid chromosome number.

16. Somatic fusion is a technique which is used to

A fuse cells from different species of animal

B fuse cells from different species of plant

C transfer genetic information into a bacterium

D alter the genes carried on a plasmid.

[Turn over

17. The graph shows the carbon dioxide gain or loss in a shade plant and in a sun plant during part of a day in summer.

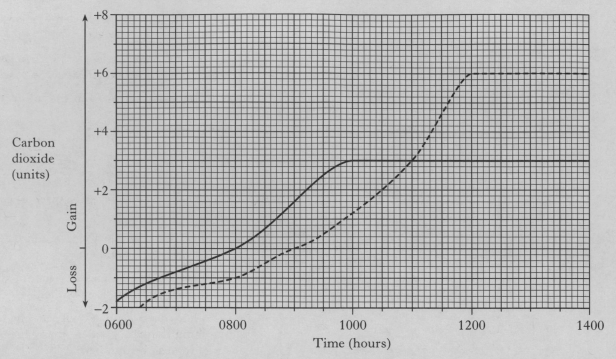

At what time does the shade plant reach compensation point?

A 0800 hours

B 0900 hours

C 1000 hours

D 1200 hours

18. The table shows water gain and loss in a plant on two consecutive days.

	Water gain (cm^3)	Water loss (cm^3)
First day	100	120
Second day	95	90

Conditions on the second day may have differed from conditions on the first day in some of the following ways.

1 Higher temperature

2 Lower windspeed

3 Lower humidity

4 Lower temperature

Which two conditions could account for the differences in water gain and loss from the first day to the second day?

A 1 and 2

B 1 and 3

C 2 and 4

D 3 and 4

19. Grass can survive despite being grazed by herbivores such as sheep and cattle. It is able to tolerate grazing because it

A is a wind-pollinated plant

B grows constantly throughout the year

C possesses poisons which protect it from being eaten entirely

D has very low growing points which send up new leaves when older ones are eaten.

20. When the intensity of grazing by herbivores increases in a grassland ecosystem, diversity of plant species may increase as a result.

Which statement explains this observation?

A Few herbivores are able to graze on every plant species present.

B Grazing stimulates growth in some plant species.

C Vigorous plant species are grazed so weaker competitors can also thrive.

D Plant species with defences against grazing are selected.

21. Which of the following describes an advantage of habituation to an animal?

A The animal becomes very good at an action which is performed repeatedly.

B An animal shows the same behaviour patterns as all those of the same species.

C A particular response is learned very quickly.

D Energy is not wasted in responding to harmless stimuli.

22. Which of the following examples of bird behaviour would result in reduced interspecific competition?

A Great Tits with the widest stripe on their breast feed first when food is scarce.

B Sooty Terns feed on larger fish than other species of tern which live in the same area.

C Pelicans searching for food form a large circle round a shoal of fish, then dip their beaks into the water simultaneously.

D Predatory gulls have difficulty picking out an individual puffin from a large flock.

23. The table shows the relative percentages by mass of the major chemical groups in a sample of human tissue.

The remaining percentage is made up of water.

Chemical group	%
Carbohydrate	5
Protein	18
Lipid	10
Other organic material	2
Inorganic material	1

What mass of water is present in a 250 g sample of this tissue?

A   64 g

B   36 g

C   90 g

D   160 g

[Turn over

24. The diagram below shows the human body's responses to temperature change.

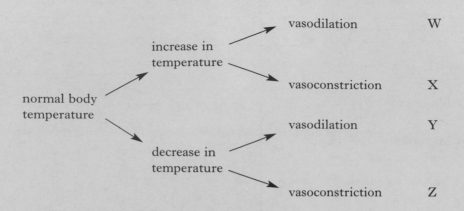

Which letters indicate negative feedback control of body temperature?

A   W and Y

B   W and Z

C   X and Y

D   X and Z

25. Muscle cells differ from nerve cells because

A   they contain different genes

B   different genes are switched on during development

C   the genetic code is different in each cell

D   they have different chromosomes.

26. A deficiency of Vitamin D in humans leads to rickets as a result of poor absorption of

A   nitrate

B   calcium

C   iron

D   phosphate.

27. Which line of the table identifies correctly the hormones which stimulate the inter-conversion of glucose and glycogen?

	glucose → glycogen	glycogen → glucose
A	insulin	glucagon and adrenaline
B	glucagon and insulin	adrenaline
C	adrenaline and glucagon	insulin
D	adrenaline	glucagon and insulin

28. Which line in the table describes body temperature in endotherms and ectotherms?

	Regulated by metabolism	Regulated by behaviour	Varies with the environmental temperature
A	ectotherm	endotherm	ectotherm
B	endotherm	ectotherm	endotherm
C	endotherm	ectotherm	ectotherm
D	ectotherm	endotherm	endotherm

29. Chlorophyll contains the metal ion

A   iron

B   copper

C   magnesium

D   calcium.

30. A species of plant was exposed to various periods of light and dark, after which the flowering response was observed.

The results are shown below.

Light period (hours)	Dark period (hours)	Response of plant
4	20	Maximum flowering
4	10	Flowering
6	18	Maximum flowering
14	10	Flowering
18	9	No flowering
18	6	No flowering
18	10	Flowering

What appears to be the critical factor which stimulates flowering?

A   A minimum dark period of 10 hours

B   A light and dark cycle of at least 14 hours

C   A maximum dark period of 10 hours

D   A dark period of at least 20 hours

Candidates are reminded that the answer sheet MUST be returned INSIDE the front cover of this answer book.

[Turn over

SECTION B

All questions in this section should be attempted.

Marks

1. Two magnified unicellular organisms are shown in the diagrams.

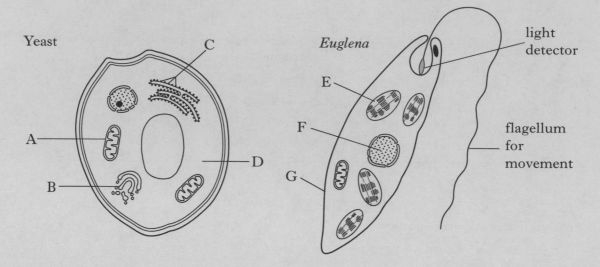

(a) (i) Name the **two** chemical components of structure G.

1 _____

2 _____

1

(ii) Complete the table by inserting letters from the diagrams to show where each process takes place.

Process	Letter
Glycolysis	
Transcription	

2

DO NOT WRITE IN THIS MARGIN

Marks

1. (continued)

(*b*) What evidence from the diagram supports the statement that yeast cells secrete enzymes?

____ **1**

(*c*) *Euglena* lives in pond water. Explain how the structure of *Euglena* shown in the diagram allows it to photosynthesise efficiently.

____

____

____ **2**

[Turn over

DO NOT
WRITE IN
THIS
MARGIN

Marks

2. An investigation was carried out into the effects of osmosis on beetroot tissue.

Pieces of beetroot were immersed in salt solutions of different concentration for one hour.

The results are shown in the table.

Concentration of salt solution (M)	Mass of beetroot at start (g)	Mass of beetroot after 1 hour (g)	Percentage change in mass (%)
0·05	4·0	4·8	+20
0·10	3·5	4·2	+20
0·20	4·4	4·7	+7
0·25	3·7	3·7	0
0·35	3·9	3·4	−13
0·40	3·5	2·8	−20

(*a*) On the grid, plot a line graph to show the percentage change in mass of the beetroot pieces against concentration of salt solution.

(Additional graph paper, if required, may be found on page 36.)

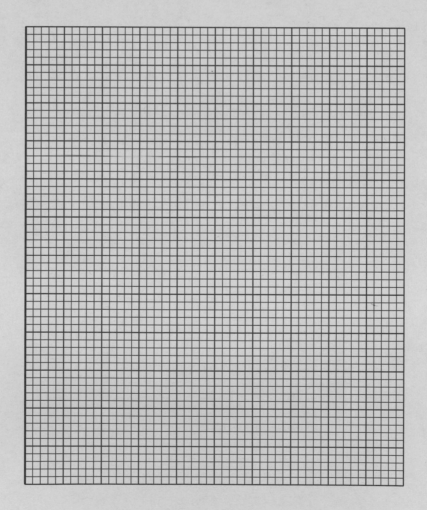

2

Marks

2. (continued)

(b) (i) Identify a concentration of salt solution used in the investigation that is hypertonic to the beetroot cell sap.

Explain your choice.

_____ M

Explanation _____

_____

_____   1

(ii) What term describes the condition of a plant cell after immersion for one hour in a 1·0 M salt solution?

_____   1

(c) From the information given, why was it good experimental practice to use percentage change in mass when comparing results?

_____

_____   1

(d) Predict the percentage change in mass of a piece of beetroot immersed in 0·45 M salt solution for one hour.

_____ %   1

(e) In setting up this investigation, variables were controlled to ensure that the results obtained would be valid.

Identify **one** variable related to the salt solutions and **one** variable related to the beetroot tissue which must be controlled.

Salt solutions _____ 1

Beetroot tissue _____ 1

[Turn over

Marks

3. (*a*) The diagram shows the role of the cytochrome system in aerobic respiration.

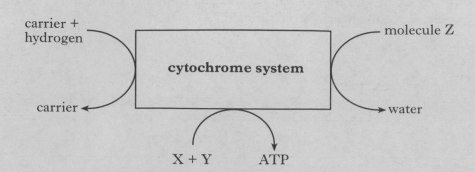

(i) State the exact location of the cytochrome system in a cell.

_____ **1**

(ii) Name the carrier that brings the hydrogen to the cytochrome system.

_____ **1**

(iii) Name molecules X, Y and Z.

X _____ Y _____ **1**

Z _____ **1**

(*b*) The graph shows the effect of different conditions on the uptake of nitrate ions by barley roots.

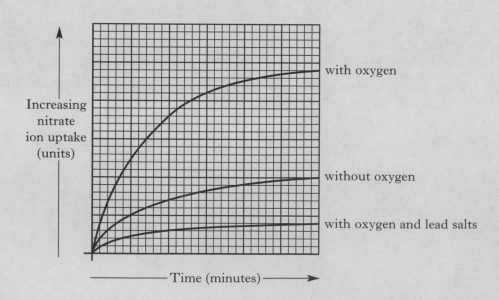

(i) State the importance of nitrate for the growth of barley plants.

_____

_____ **1**

DO NOT WRITE IN THIS MARGIN

Marks

3. (*b*) (continued)

(ii) Explain why the uptake of nitrate ions is greater when oxygen is present.

____

____

______________________________ **2**

(iii) Explain the effect of lead salts on nitrate ion uptake.

____

____

______________________________ **2**

[Turn over

Marks

4. (*a*) The replication of part of a DNA molecule is represented in the diagram.

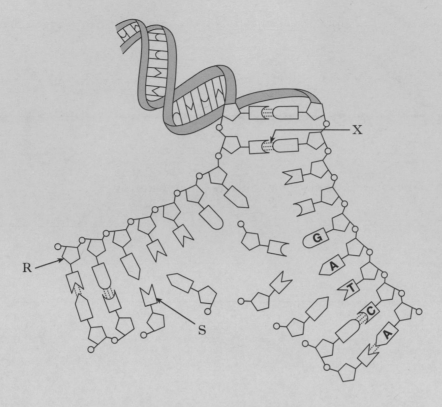

(i) Name the nucleotide component R and the base S.

R _____   1

S _____   1

(ii) Name the type of bond labelled X.

X _____   1

(*b*) Explain why DNA replication must take place before a cell divides.

_____

_____

_____   1

Marks

4. (continued)

(c) Part of one strand of a DNA molecule used to make mRNA contains the following base sequence.

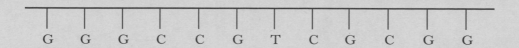

| G | G | G | C | C | G | T | C | G | C | G | G |

The table shows the names of six amino acids together with some of their mRNA codons.

Amino acid	mRNA codon (s)	
Glycine	GGG	GGC
Serine	UCG	AGC
Proline	CCG	CCC
Arginine	CGG	
Alanine	GCC	
Threonine	ACG	

(i) Use the information to give the order of amino acids coded for by the DNA base sequence.

_____   1

(ii) What name is given to a part of a DNA molecule which carries the code for making **one** protein?

_____   1

(d) Name the molecules that transport amino acids to the site of protein synthesis.

_____   1

(e) Complete the diagram below which shows information about protein classification.

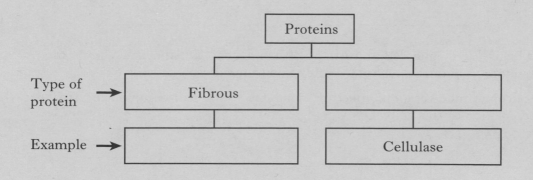

1

[Turn over

DO NOT
WRITE IN
THIS
MARGIN

Marks

5. (*a*) The diagram shows two chromosomes and their appearance after a mutation has occurred.

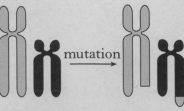

(i) Name this type of chromosome mutation.

_____ 1

(ii) Name a mutagenic agent which could have caused this mutation.

_____ 1

(*b*) Individuals with Down's Syndrome have 47 chromosomes in each cell instead of 46.

How does this change in chromosome number arise?

_____ 1

(*c*) The diagram shows part of the normal amino acid sequence of an enzyme involved in a metabolic pathway. It also shows the altered sequence obtained after a gene mutation had occurred.

Normal amino
acid sequence

Altered amino
acid sequence

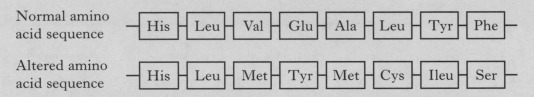

(i) Name a type of gene mutation which could have produced this altered amino acid sequence.

_____ 1

(ii) Explain the effect this gene mutation would have on the metabolic pathway in which this enzyme is involved.

_____

_____ 1

(*d*) The DNA in one cell consists of 40 000 genes. During DNA replication, random mutations occur at the rate of one altered gene in every 625.

Calculate the average number of mutations which will occur during the full replication of this cell's DNA.

Space for working

_____ 1

Marks

6. (*a*) A gene from a jellyfish can be inserted into a bacterial plasmid using a genetic engineering procedure.

Some of the stages involved are shown in the diagram.

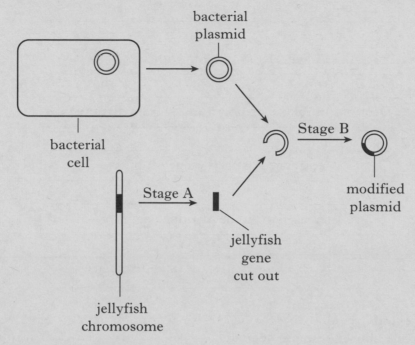

(i) Give **one** method which could be used for locating the gene in the jellyfish chromosome.

_____   1

(ii) Name the enzymes involved in the following stages of the genetic engineering procedure.

1 Cutting the jellyfish gene out of its chromosome (Stage A).

_____   1

2 Sealing the jellyfish gene into the bacterial plasmid (Stage B).

_____   1

(*b*) Name **one** human hormone that is manufactured by genetically engineered bacteria.

_____   1

[Turn over

DO NOT
WRITE IN
THIS
MARGIN

Marks

7. (*a*) The flow diagram shows some stages in the regulation of water concentration of blood in mammals.

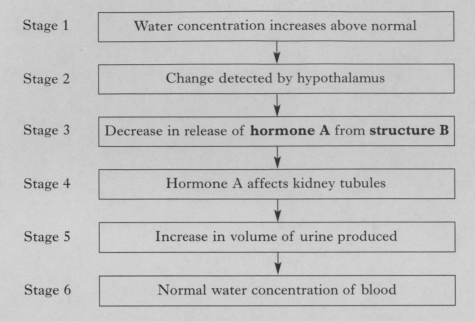

Stage 1 — Water concentration increases above normal

Stage 2 — Change detected by hypothalamus

Stage 3 — Decrease in release of **hormone A** from **structure B**

Stage 4 — Hormone A affects kidney tubules

Stage 5 — Increase in volume of urine produced

Stage 6 — Normal water concentration of blood

(i) Give **one** reason why the water concentration of the blood could increase above normal at stage 1.

_____ 1

(ii) 1 Name hormone A _____ 1

2 Name structure B _____ 1

(iii) Describe how hormone A is transported to the kidney tubules.

_____ 1

(iv) Describe the effect of hormone A on the kidney tubules.

_____

_____ 1

(v) What would be the effect of a decrease in hormone A on the concentration of salts in the urine?

_____ 1

DO NOT WRITE IN THIS MARGIN

Marks

7. (continued)

(*b*) Salmon migrate between sea water and fresh water.

The table contains statements about osmoregulation in salmon.

For each statement, tick (✓) **one** box to show whether the statement is true for a salmon living in sea water or in fresh water.

Statement	*Sea water*	*Fresh water*
Salmon drinks a large volume of water		
Salmon produces a large volume of urine		
Chloride secretory cells pump out ions		
Salmon gains water by osmosis		

2

[Turn over

Marks

8. An experiment was carried out to investigate the growth of pea plants kept in a high light intensity following germination.

The graph shows the average dry mass and average shoot length of the pea plants.

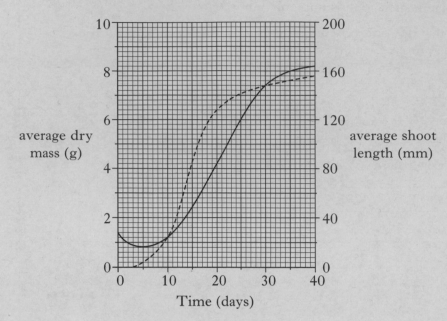

KEY

average shoot
length - - - - -

average dry
mass ————

(a) (i) From the graph, how many days does it take for the shoot to emerge from the seed?

_____ days   **1**

(ii) During which 5 day period is there the greatest increase in average shoot length? Tick (✓) one box.

Day 5–10	Day 10–15	Day 15–20	Day 20–25	Day 25–30

1

(iii) Explain the changes in average dry mass of the plants during the first fifteen days.

_____

_____

_____

2

(iv) Explain why measurement of average shoot length alone may not provide a reliable estimate of plant growth.

_____

_____

_____

1

8. **(a)** **(continued)**

Marks

(v) On day 30 the shoots made up 50% of the average dry mass of the plants. Calculate the average dry mass of the shoots per millimetre.

Space for calculation

_____ g per mm   1

(b) The experiment was repeated with pea plants kept in the dark.

Complete the following to show how the results on day 15 would compare with the results obtained from plants grown in the light.

In each case, underline **one** alternative and give a reason to justify your choice.

(i) Average dry mass would be $\left\{ \begin{array}{l} \text{greater.} \\ \text{less.} \\ \text{the same.} \end{array} \right\}$

Reason _____   1

(ii) Average shoot length would be $\left\{ \begin{array}{l} \text{greater.} \\ \text{less.} \\ \text{the same.} \end{array} \right\}$

Reason _____   1

(c) The grid shows some of the effects of the plant growth substances Indole Acetic Acid (IAA) and Gibberellic Acid (GA) on the growth and development of plants.

A	stimulates α-amylase production in barley grains	B	promotes the formation of fruit	C	inhibits leaf abscission
D	causes apical dominance	E	involved in phototropism	F	breaks dormancy of buds

(i) Use **all** the letters from the grid to complete the table to show which effects are caused by IAA and which are caused by GA.

Effects caused by IAA	*Effects caused by GA*

3

(ii) Give **one** practical application of plant growth substances.

_____

_____   1

Marks

9. The diagram represents a section through a woody twig with an area enlarged to show the xylem vessels present.

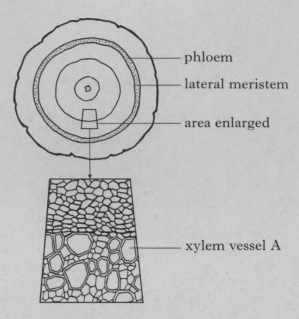

phloem

lateral meristem

area enlarged

xylem vessel A

(a) Name the lateral meristem shown in the diagram.

_____ 1

(b) Explain how the appearance of xylem vessel A indicates that it was formed in the spring.

_____

_____

_____ 1

(c) What name is given to the area of the woody twig section that represents the xylem tissue growth occurring in one year?

_____ 1

Marks

10. Duckweed (*Lemna*) is a hydrophyte that has leaf-like structures which float on the surface of pondwater.

Some *Lemna* plants are shown in the diagram together with a magnified vertical section through one of the floating leaf-like structures.

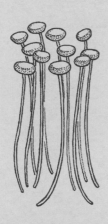

Lemna plants

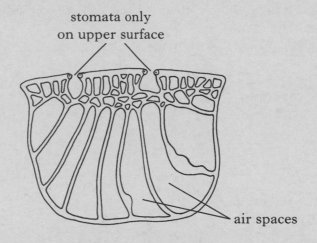

magnified vertical section

(*a*) Complete the table to describe the effect of each adaptation in *Lemna*.

Adaptation	Effect
Many large air spaces	
Stomata on upper surface	

1

1

(*b*) What term describes a plant that is adapted to live in a hot, dry habitat?

_____ 1

[Turn over

Marks

11. An investigation was carried out into the effects of competition when two species of flour beetle, *Tribolium confusum* and *Tribolium castaneum*, were kept together in a container with a limited food supply.

Tribolium beetles can be infected by a parasite which causes disease.

Graph 1 shows the numbers of the two species over the period of time in the absence of the parasite.

Graph 2 shows the effect of the presence of the parasite on the beetle numbers.

Graph 1 Parasite absent

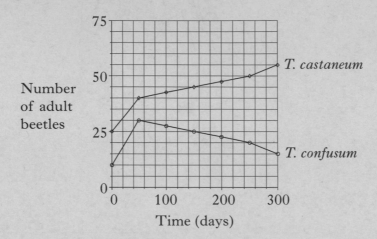

Graph 2 Parasite present

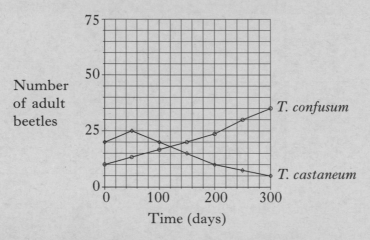

(a) Use values from **Graph 1** to describe how the numbers of *T. confusum* change over the period of the investigation.

_____

_____

_____

2

Marks

11. **(continued)**

(*b*) From **Graph 1**, express as the simplest whole number ratio the population size of *T. confusum* to *T. castaneum* at 250 days.

Space for calculation

T. confusum : *T. castaneum* _____ : _____   1

(*c*) From **Graph 2**, calculate the percentage increase in the *T. confusum* population over the 300 days of the investigation.

Space for calculation

_____ % increase   1

(*d*) Suggest an explanation for the improved growth of the *T. confusum* population in the presence of the parasite.

_____

_____

_____   2

(*e*) From the information in the graphs, suggest an improvement to the design of the investigation.

_____

_____   1

(*f*) Certain factors may affect the numbers of beetles in this investigation.

Place ticks in the table to show whether each factor would have a density-dependent effect or a density-independent effect.

Factors	*Density-dependent*	*Density-independent*
Presence of disease causing parasites		
Availability of food		
Extreme temperature		

2

Page twenty-seven   **[Turn over**

Marks

12. The diagram shows parts of the chromosome in the bacterium *E. coli*. The list has three molecules involved in the genetic control of lactose metabolism.

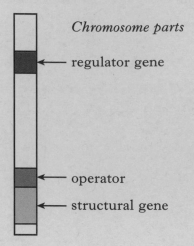

Chromosome parts

← regulator gene

← operator

← structural gene

List of molecules

lactose-digesting enzyme
repressor
inducer

(*a*) Complete the table by writing **True** or **False** in each of the spaces provided.

Statement	True/False
The repressor can bind to the operator.	
The structural gene codes for the repressor.	
The inducer can bind to the repressor.	
The regulator gene codes for the lactose-digesting enzyme.	

2

(*b*) Name the inducer molecule.

_____

1

(*c*) Give **one** advantage to *E. coli* of having this type of genetic control system.

_____

_____

1

SECTION C

Both questions in this section should be attempted.

Note that each question contains a choice.

Questions 1 and 2 should be attempted on the blank pages which follow.

Supplementary sheets, if required, may be obtained from the invigilator.

Labelled diagrams may be used where appropriate.

Marks

1. Answer **either** A **or** B.

    A.   Give an account of meiosis under the following headings:

       (i)   first meiotic division;   **6**

       (ii)   second meiotic division;   **2**

       (iii)   importance of meiosis.   **2**

           (10)

    OR

    B.   Give an account of the evolution of new species under the following headings:

       (i)   isolating mechanisms;   **4**

       (ii)   effects of mutations and natural selection.   **6**

           (10)

In question 2, ONE mark is available for coherence and ONE mark is available for relevance.

2. Answer **either** A **or** B.

    A.   Give an account of chloroplast structure in relation to the location of the stages of photosynthesis and describe the separation of photosynthetic pigments by chromatography.   **(10)**

    OR

    B.   Give an account of the nature of viruses and the production of more viruses.   **(10)**

[END OF QUESTION PAPER]

DO NOT
WRITE IN
THIS
MARGIN

SPACE FOR ANSWERS

[BLANK PAGE]

FOR OFFICIAL USE

Total for
Sections
B and C

X007/301

NATIONAL
QUALIFICATIONS
2005

WEDNESDAY, 18 MAY
1.00 PM – 3.30 PM

BIOLOGY
HIGHER

Fill in these boxes and read what is printed below.

Full name of centre

Town

Forename(s)

Surname

Date of birth
Day Month Year   Scottish candidate number   Number of seat

SECTION A—Questions 1–30 (30 marks)

Instructions for completion of Section A are given on page two.

SECTIONS B AND C (100 marks)

1   (a)   All questions should be attempted.

(b)   It should be noted that in **Section C** questions 1 and 2 each contain a choice.

2   The questions may be answered in any order but all answers are to be written in the spaces provided in this answer book, and must be written clearly and legibly in ink.

3   Additional space for answers and rough work will be found at the end of the book. If further space is required, supplementary sheets may be obtained from the invigilator and should be inserted inside the **front** cover of this book.

4   The numbers of questions must be clearly inserted with any answers written in the additional space.

5   Rough work, if any should be necessary, should be written in this book and then scored through when the fair copy has been written. If further space is required a supplementary sheet for rough work may be obtained from the invigilator.

6   Before leaving the examination room you must give this book to the invigilator. If you do not, you may lose all the marks for this paper.

SCOTTISH
QUALIFICATIONS
AUTHORITY

Read carefully

1   Check that the answer sheet provided is for **Biology Higher (Section A)**.

2   Check that the answer sheet you have been given has **your name**, **date of birth**, **SCN** (Scottish Candidate Number) and **Centre Name** printed on it.

Do not change any of these details.

3   If any of this information is wrong, tell the Invigilator immediately.

4   If this information is correct, **print** your name and seat number in the boxes provided.

5   Use **black** or **blue ink** for your answers. **Do not use red ink**.

6   The answer to each question is **either** A, B, C or D. Decide what your answer is, then put a horizontal line in the space provided (see sample question below).

7   There is **only one correct** answer to each question.

8   Any rough working should be done on the question paper or the rough working sheet, **not** on your answer sheet.

9   At the end of the exam, put the **answer sheet for Section A inside the front cover of this answer book**.

Sample Question

The apparatus used to determine the energy stored in a foodstuff is a

A   respirometer

B   calorimeter

C   klinostat

D   gas burette

The correct answer is **B**—calorimeter. The answer **B** has been clearly marked with a horizontal line (see below).

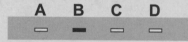

Changing an answer

If you decide to change your answer, cancel your first answer by putting a cross through it (see below) and fill in the answer you want. The answer below has been changed to **B**.

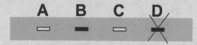

If you then decide to change back to an answer you have already scored out, put a tick (✓) to the **right** of the answer you want, as shown below:

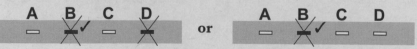

SECTION A

All questions in this section should be attempted.

Answers should be given on the separate answer sheet provided.

1. When a red blood cell is immersed in a hypertonic solution it will

   A   shrink

   B   become flaccid

   C   burst

   D   become turgid.

2. The diagram below represents some of the structures present in a plant cell.

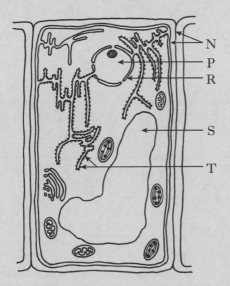

   Which line in the table matches the structures with the materials of which they are mainly composed?

	Materials	
	protein and phospholipid	nucleic acid and protein
A	R	P
B	R	S
C	T	R
D	R	N

3. Which of the following is a structural carbohydrate?

   A   Glucose

   B   Starch

   C   Glycogen

   D   Cellulose

4. The graph illustrates the effects of light intensity, temperature and carbon dioxide (CO_2) concentration on the rate of photosynthesis.

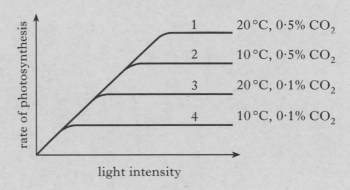

   Which of the following pairs of lines in the graph suggest that carbon dioxide is acting as a limiting factor?

   A   2 and 4

   B   2 and 3

   C   1 and 4

   D   1 and 2

5. Which of the following elements is essential to the formation of chlorophyll?

   A   Potassium

   B   Magnesium

   C   Copper

   D   Calcium

6. Which of the following is composed of protein?

   A   Nucleotide

   B   Glycogen

   C   Antibody

   D   Polysaccharide

[Turn over

7. How many adenine molecules are present in a DNA molecule of 2000 bases, if 20% of the base molecules are cytosine?

   A 200

   B 300

   C 400

   D 600

8. Which of the following statements is true of all viruses?

   A They have a protein-lipid coat and contain DNA.

   B They have a protein-lipid coat and contain RNA.

   C They have a protein coat and a nucleus.

   D They have a protein coat and contain nucleic acid.

9. The genes of viruses are composed of

   A either DNA or RNA

   B DNA only

   C RNA only

   D enzymes and nucleic acids.

10. In infertility clinics, samples of semen are collected for testing.

    The table below refers to the analysis of semen samples taken from five men.

Semen sample	1	2	3	4	5
Number of sperm in sample (millions/cm^3)	40	19	25	45	90
Active sperm (percent)	65	60	75	10	70
Abnormal sperm (percent)	30	20	90	30	10

A man is fertile if his semen contains at least 20 million sperm cells/cm^3 and at least 60% of the sperm cells are active and at least 60% of the sperm cells are normal.

The semen samples that were taken from infertile men are

   A samples 3 and 4 only

   B samples 2 and 4 only

   C samples 2, 3 and 4 only

   D samples 1, 2, 4 and 5 only.

11. Alleles can be described as

    A opposite types of gamete

    B different versions of a gene

    C identical chromatids

    D non-homologous chromosomes.

12. Which of the following defines linkage?

    A Genes which are transferred from one chromosome pair to another

    B Genes which are present on the same chromosome

    C Genes which are transferred from one chromosome to its partner

    D Genes which are present on different chromosomes

13. The table below shows the recombination frequency between genes on a chromosome.

Crossing over between genes	Recombination frequency
F and G	4%
F and J	6%
G and H	6%
H and J	4%

Use the information in the table to work out the order of genes on the chromosome.

The order of the genes is

    A   H   G   F   J

    B   F   G   H   J

    C   F   G   J   H

    D   G   H   F   J.

14. In *Drosophila*, white eye colour is a sex-linked recessive character. If a homozygous white-eyed female is crossed with a red-eyed male, what will be the phenotype of the first generation?

    A All females will be white-eyed and all males red-eyed.

    B All females will be red-eyed and all males white-eyed.

    C All females will be red-eyed and 1 in 2 males will be white-eyed.

    D 1 in 4 will be white-eyed irrespective of sex.

15. Which of the following may result in the presence of an extra chromosome in the cells of a human being?

    A Non-disjunction

    B Crossing over

    C Segregation

    D Inversion

16. Which of the following is an example of the result of natural selection?

    A Modern varieties of potato have been produced from wild varieties.

    B Ayrshire cows have been selected through breeding for milk production.

    C Bacterial species have developed resistance to antibiotics.

    D Varieties of tomato plants have resistance to fungal diseases through somatic fusion.

17. The dark variety of the peppered moth became common in industrial areas of Britain following the increase in the production of soot during the Industrial Revolution.

The increase in the dark form was due to

    A dark moths migrating to areas which gave the best camouflage

    B a change in the prey species taken by birds

    C an increase in the mutation rate

    D a change in selection pressure.

18. Which of the following is true of the kidneys of a salt-water bony fish?

    A They have few large glomeruli.

    B They have few small glomeruli.

    C They have many large glomeruli.

    D They have many small glomeruli.

19. The Soft Brome Grass and Long Beaked Storksbill are species of plant which grow on the grasslands of California. The Storksbill is a low-growing plant with a more extensive root system than the Soft Brome, but does not grow as tall as the Soft Brome.

Under which of the following conditions would the Storksbill become the more abundant species?

    A Drought

    B High soil moisture levels

    C High light intensity

    D Shade

20. Which of the following best describes habituation?

    A The same escape response is performed repeatedly.

    B The same response is always given to the same stimulus.

    C A harmless stimulus ceases to produce a response.

    D Behaviour is reinforced by regular repetition.

[Turn over

21. Hawks are predators which attack flocks of pigeons. The graph below shows how attack success by a hawk varies with the number of pigeons in a flock.

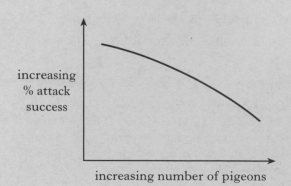

increasing number of pigeons

Which of the following statements could explain the observations shown in the graph?

A A hawk only needs to eat a small percentage of a large flock of prey.

B Co-operative hunting is more effective with small numbers of prey.

C A predator can be more selective when prey numbers increase.

D A hawk has difficulty focussing on one pigeon in a large flock.

22. Root tips are widely used for the study of mitosis because

A the cells are larger than other cells

B they contain many meristematic cells

C their nuclei have large chromosomes

D their nuclei stain easily.

23. The graphs below show the average yearly increase in height of girls and boys.

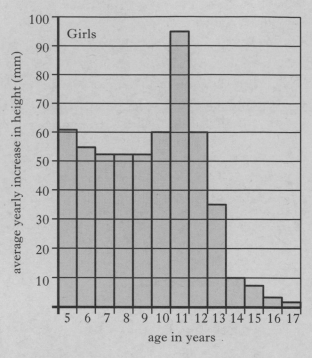

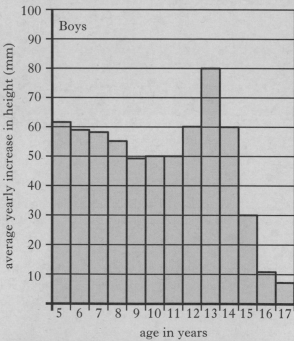

Which of the following statements is correct?

A The greatest average yearly increase for boys occurs one year later than the greatest average yearly increase for girls.

B Boys are still growing at seventeen but girls have stopped growing by this age.

C Between the ages of five and eight boys grow more than girls.

D There is no age when boys and girls show the same average yearly increase in height.

24. The following diagram shows an enzyme-controlled metabolic pathway.

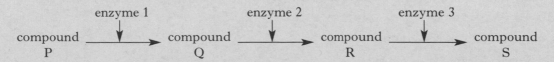

compound P → enzyme 1 → compound Q → enzyme 2 → compound R → enzyme 3 → compound S

If enzyme 2 is inactivated (eg by adding an inhibitor) at time X shown in the graphs below, which graph predicts correctly the final concentration of compounds Q and R?

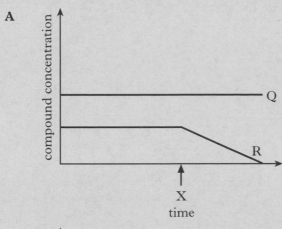

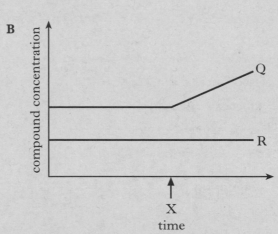

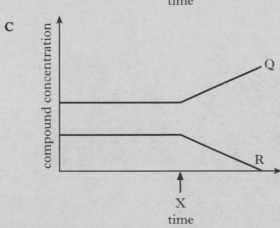

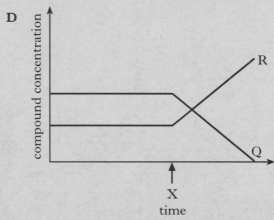

25. The table shows the results of an experiment carried out to study the effects of a plant growth substance on the roots of tomato plants.

Concentration of growth substance		Control 0 mg/litre	10^{-1} mg/litre
Average length of 20 roots	Before treatment	16 mm	16 mm
	After treatment	24 mm	20 mm

Which of the following states the effect of the plant growth substance on the lengths of the roots compared to the control treatment?

A   25 percent stimulation

B   50 percent stimulation

C   25 percent inhibition

D   50 percent inhibition

26. A short day plant is one which

A   will flower only if the night length is less than the critical value

B   will flower only if daylight is less than 12 hours

C   will flower only if the hours of daylight are less than a critical value

D   flowers only if the hours of daylight are more than a critical value.

27. A plant becomes etiolated when it

A   grows in poor soil

B   grows in the dark

C   is treated with gibberellin

D   has the apical bud removed.

28. If the concentration of glucose in the blood of a healthy man or woman rises above normal, the pancreas produces

   A more insulin but less glucagon

   B more insulin and more glucagon

   C less insulin but more glucagon

   D less insulin and less glucagon.

29. If body temperature drops below normal, which of the following would result?

   A Vasodilation of skin capillaries

   B Vasoconstriction of skin capillaries

   C Decreased metabolic rate

   D Increased sweating

30. The diagram below represents a sandy coastal area. The sand deposits support various communities of plants.

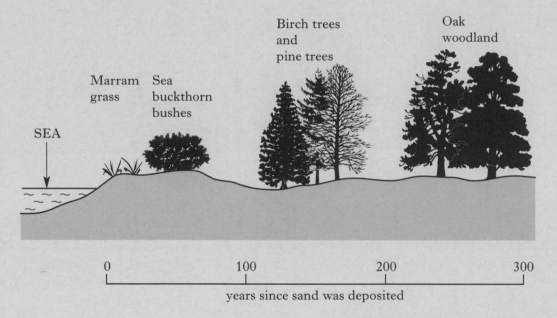

What term is used to describe the sequence of communities shown?

   A Colonisation

   B Climax

   C Progression

   D Succession

Candidates are reminded that the answer sheet MUST be returned INSIDE the front cover of this answer book.

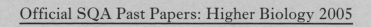

[Turn over for Section B on *Page ten*

Marks

SECTION B

All questions in this section should be attempted.

1.   The diagram shows a mitochondrion from a human muscle cell.

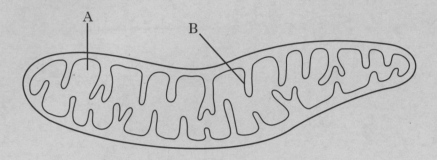

(*a*)   Name regions A and B.

A _____

B _____   1

(*b*)   The table shows some substances involved in respiration.

(i)   Complete the table by inserting the number of carbon atoms present in each substance.

Substance	Number of carbon atoms present
Pyruvic acid	
Acetyl group	
Citric acid	

2

(ii)   To which substance is the acetyl group attached before it enters the citric acid cycle?

_____   1

1.   (continued)

Marks

(c)   In region B, hydrogen is passed through a series of carriers in the cytochrome system as shown in the diagram below.

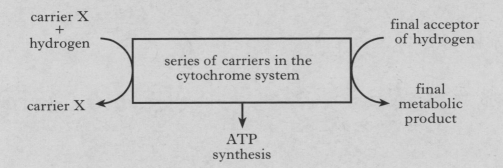

(i)   Name carrier X.

_____   1

(ii)   Name the final acceptor of hydrogen.

_____   1

(iii)   Describe the importance of ATP in cells.

_____

_____   1

(iv)   The quantity of ATP present in the human body remains relatively constant yet ATP is continually being broken down.

Suggest an explanation for this observation.

_____

_____

_____   1

(d)   Name the final metabolic product of **anaerobic** respiration in a muscle cell.

_____   1

[Turn over

Marks

2. The diagram shows two different types of blood cell involved in the defence of the human body.

phagocyte   cell X

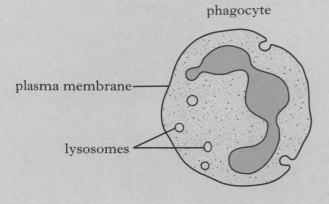
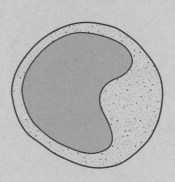

plasma membrane

lysosomes

(a) Describe how the plasma membrane and the lysosomes of phagocytes are involved in helping to destroy bacteria.

   (i) plasma membrane _____

_____ 1

  (ii) lysosomes _____

_____ 1

(b)   (i) Name cell X.

_____ 1

  (ii) Explain how cell X may be involved in tissue rejection following a transplant operation.

_____

_____

_____ 1

 (iii) What treatment is given to prevent tissue rejection?

_____ 1

Marks

3. In Dachshund dogs, the genes for hair texture and hair length are located on different chromosomes.

The allele for wire hair (**A**) is dominant to the allele for smooth hair (**a**).
The allele for short hair (**B**) is dominant to the allele for long hair (**b**).

Wire hair is **always** short so dogs with allele **A** are **always** short haired.

Two Dachshunds with the genotype **AaBb** were crossed.

(*a*) State the phenotype of the parents in this cross.

_____   1

(*b*) The grid shows all the genotypes of the offspring that may arise from this cross.

Complete the grid by adding the genotypes of the male and female gametes.

Male gametes

	¹ **AABB**	**AABb**	**AaBB**	**AaBb**
	AABb	² **AAbb**	**AaBb**	**Aabb**
	AaBB	**AaBb**	³ **aaBB**	**aaBb**
	AaBb	**Aabb**	**aaBb**	⁴ **aabb**

Female gametes

1

(*c*) Complete the table below to give the phenotypes of the offspring indicated by the shaded boxes numbered 1 to 4 on the grid.

Box	*Phenotype*
1	
2	
3	
4	

2

(*d*) From the grid, calculate the expected ratio of the phenotypes of **all** the offspring from this cross.

Space for working

_____ wire short hair : _____ smooth short hair : _____ smooth long hair   1

Marks

4. (*a*) The diagram shows the amino acid sequences of a fish hormone and two human hormones which may have evolved from it.

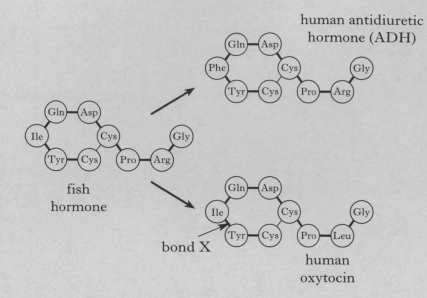

Amino acid key

Arg	arginine
Asp	aspartic acid
Cys	cysteine
Gln	glutamine
Gly	glycine
Ile	isoleucine
Leu	leucine
Phe	phenylalanine
Pro	proline
Tyr	tyrosine

(i) Name the type of bond represented by X.

_____ 1

(ii) In the evolution of human oxytocin from the fish hormone, a gene mutation resulted in the amino acid arginine being replaced by leucine.

The table shows four of the mRNA codons for the amino acids arginine and leucine.

Codons for arginine	Codons for leucine
CGU	CUU
CGC	CUC
CGA	CUA
CGG	CUG

Name the type of gene mutation that occurred and justify your answer.

Type of gene mutation _____ 1

Justification _____

_____ 1

(iii) Describe the change in protein structure that occurred in the evolution of human antidiuretic hormone (ADH) from the fish hormone.

_____

_____ 1

Marks

4. **(continued)**

   (*b*) Antidiuretic hormone (ADH) is involved in osmoregulation in humans.

   (i) Name the gland that releases ADH.

   _____

   1

   (ii) The graphs show the effects of increasing blood solute concentration and increasing blood volume on the plasma ADH concentration.

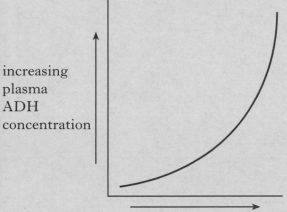

increasing plasma ADH concentration

increasing blood solute concentration

increasing plasma ADH concentration

increasing blood volume

Use the information in the graphs to complete the table by using the terms "increases", "decreases" or "stays the same" to show the effect of various activities on the plasma ADH concentration.

Each term may be used **once**, **more than once** or **not at all**.

Activity	*Effect on plasma ADH concentration*
Drinking fresh water	
Sweating	
Eating salty food	
Severe bleeding	

2

   (iii) Describe the effect that an increase in plasma ADH concentration has on the activity of kidney tubules.

   _____

   _____

   1

[Turn over

Marks

5. The diagram shows how an isolating mechanism can divide a population of one species into two sub-populations and then act as a barrier to prevent gene exchange between them.

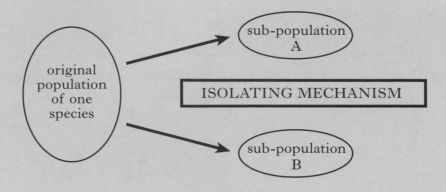

(a) Name **one** type of isolation that could prevent gene exchange between the two sub-populations.

_____   1

(b) Over a long period of time, the gene pools of sub-populations A and B become different from each other.

   (i) Explain how mutations and natural selection account for the differences.

     1. Mutations _____

     _____

     _____   1

     2. Natural selection _____

     _____

     _____

     _____   2

   (ii) Eventually, sub-populations A and B may become two different species. What evidence would confirm that this had happened?

_____

_____   1

Marks

6. The list below contains terms related to genetic engineering and somatic fusion.

List of terms:

  cellulase
  gene probe
  ligase
  plasmid
  protoplast
  restriction endonuclease.

(*a*) Complete the table to match **each** of the following descriptions to the correct term from the above list.

Description	Term
Contains bacterial genes	
Cuts DNA into fragments	
Locates specific genes	
Removes plant cell walls	

2

(*b*) State the problem in plant breeding that is overcome by using the technique of somatic fusion.

1

[Turn over

Marks

7. (*a*) The diagram represents a plant with two regions magnified to show tissues involved in transport.

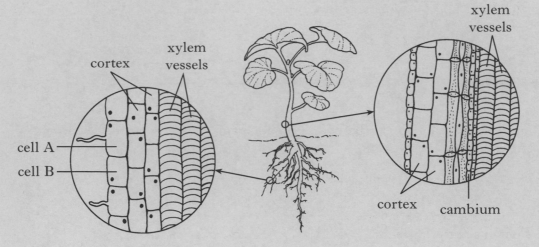

(i) Describe the process by which water moves into cell A.

_____

_____

_____   1

(ii) Cells A and B have a similar function.
Explain how the structure of cell A makes it better adapted to its function than cell B.

_____

_____

_____   1

(iii) Name the force that holds water molecules together as they travel up the xylem vessels.

_____   1

(iv) Cell division in the cambium produces new cells which then elongate and develop vacuoles.

Describe **two** further changes that take place in these cells as they differentiate into xylem vessels.

1 _____

_____   1

2 _____

_____   1

Marks

7. **(continued)**

(b) The diagrams show stomata in the lower epidermis of a leaf.

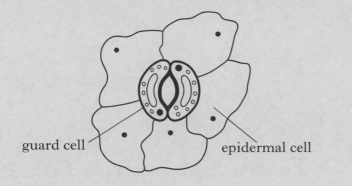

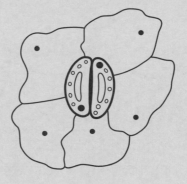

guard cell — — epidermal cell

open stoma   *closed stoma*

(i) In the following sentence, **underline** one of the alternatives in each pair to make the sentence correct.

Stomata close when water moves $\begin{Bmatrix} \text{into} \\ \text{out of} \end{Bmatrix}$ the guard cells

and they become $\begin{Bmatrix} \text{more} \\ \text{less} \end{Bmatrix}$ turgid.

1

(ii) What is the advantage to plants in having their stomata closed at night?

_____

1

(c) The grid shows factors affecting the rate of transpiration from leaves.

A increased temperature	B increased wind speed	C increased humidity
D decreased temperature	E decreased wind speed	F decreased humidity

(i) Which **three** letters indicate the changes that would result in a decrease in the rate of transpiration?

Letters _____ , _____ and _____ .

1

(ii) The transpiration stream supplies plant cells with water for photosynthesis.

Give **one** other benefit to plants of the transpiration stream.

_____

1

8. **Figure 1** shows how glycerate phosphate (GP) and ribulose bisphosphate (RuBP) are involved in the Calvin cycle.

Figure 1

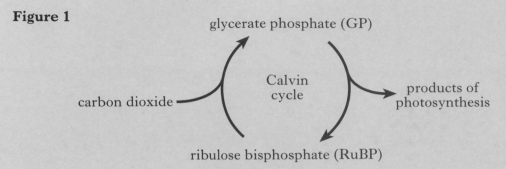

An investigation of the Calvin cycle was carried out in *Chlorella*, a unicellular alga.

Graph 1 shows the concentrations of GP and RuBP in *Chlorella* cells kept in an illuminated flask at 15 °C. The concentration of carbon dioxide in the flask was 0·05% for the first three minutes, then it was reduced to 0·005%.

Graph 1

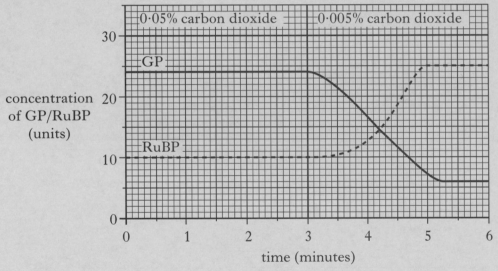

Graph 2 shows the rate of carbon dioxide fixation by *Chlorella* cells at various carbon dioxide concentrations.

Graph 2

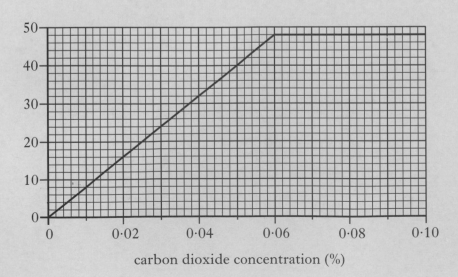

8. **(continued)**

Marks

(a) (i) Use values from **Graph 1** to describe the changes in the RuBP concentration over the first six minutes.

_____

_____

_____ 2

(ii) Use the information in **Figure 1** to explain the increase in RuBP concentration shown in **Graph 1** when the carbon dioxide concentration is decreased.

_____

_____

_____ 2

(b) From **Graph 1**, calculate the percentage decrease in the concentration of GP from three to six minutes.

Space for calculation

_____ % 1

(c) Use the terms "increase", "decrease" or "stay the same" to complete the sentence below. Each term may be used **once**, **more than once** or **not at all**.

If the carbon dioxide concentration was returned to 0·05% at 6 minutes,

the concentration of RuBP would _____

and the concentration of GP would _____ . 1

(d) From **Graph 2**, state the rate of carbon dioxide fixation by *Chlorella* at a carbon dioxide concentration of 0·01%.

_____ mmol h^{-1} 1

(e) How many times greater is the rate of carbon dioxide fixation from 0 to 3 minutes compared with 3 to 6 minutes?

Space for calculation

_____ times 1

Marks

9. African wild dogs are social animals that hunt in packs. They rely on stamina to catch grazing prey such as wildebeest.

The table shows the effect of wildebeest age on the average duration of successful chases and the percentage hunting success.

Wildebeest age	Stage	Average duration of successful chases (s)	Hunting success (%)
up to 1 year	calves	20	75
from 1 – 2 years	juveniles	120	50
over 2 years	adults	180	45

(a) Describe the effect of wildebeest age on the average duration of successful chases.

_____

_____   1

(b) How many times longer does it take the wild dogs on average to successfully hunt adult wildebeest rather than calves?

Space for working

_____ times   1

(c) Suggest a reason why hunting success is greatest with calves.

_____

_____   1

(d) Wild dogs kill a greater number of adult wildebeest than calves.

Explain this observation in terms of the economics of foraging behaviour.

_____

_____

_____   1

(e) State an advantage of cooperative hunting to the wild dogs.

_____   1

(f) Following a successful hunt, wild dogs may be displaced from their kill by spotted hyenas. What type of competition does this show?

_____   1

Marks

10. (*a*) Fulmars and Common Terns are seabirds that breed in large social groups. The table compares features of breeding in these birds.

Feature of breeding	Fulmar	Common Tern
nest distribution and location	crowded on cliff ledges	scattered on pebble beaches
egg number and colour	single white egg	three speckled eggs
chick behaviour	remains in nest until able to fly	can move short distances from nest soon after hatching

(i) Use information in the table to explain why Fulmars are less vulnerable to predation than Common Terns.

_____

_____

_____   1

(ii) Suggest how features of Common Tern eggs and chicks may increase their survival chances.

1   Eggs _____

_____

_____   1

2   Chicks _____

_____

_____   1

(*b*) Explain how living in large social groups may help animals in defence against predators.

_____

_____

_____   1

[Turn over

Marks

11. An investigation was carried out into the effect of lead ethanoate and calcium ethanoate on the activity of catalase.

Catalase is an enzyme found in yeast cells. It acts on hydrogen peroxide to produce oxygen gas.

The stages in the investigation are outlined below.

1   Three yeast suspensions were made by adding 100 mg of dried yeast to each of the following.

- $25\,cm^3$ of $0 \cdot 1$ M lead ethanoate solution
- $25\,cm^3$ of $0 \cdot 1$ M calcium ethanoate solution
- $25\,cm^3$ of water

2   The suspensions were stirred and left for 15 minutes.

3   Separate syringes were used to add $2\,cm^3$ of each yeast suspension to $10\,cm^3$ of hydrogen peroxide in 3 identical containers.

4   The volume of oxygen produced in each container was measured at 10 second intervals.

The results are shown in the table.

Time (s)	Volume of oxygen produced (cm^3)		
	yeast suspension + lead ethanoate	*yeast suspension + calcium ethanoate*	*yeast suspension + water*
0	0	0	0
10	6	32	38
20	10	62	56
30	14	74	78
40	15	88	86
50	16	90	88
60	17	90	90

(*a*)   Why was it good experimental procedure to leave the yeast suspensions for 15 minutes at stage 2?

_____

_____   1

(*b*)   Why was a separate syringe used for each yeast suspension at stage 3?

_____

_____   1

(*c*)   Identify **one** variable, not already described, that should be kept constant.

_____   1

11. (continued)

(d) The results for the yeast suspensions in 0·1 M calcium ethanoate and in water are shown in the graph.

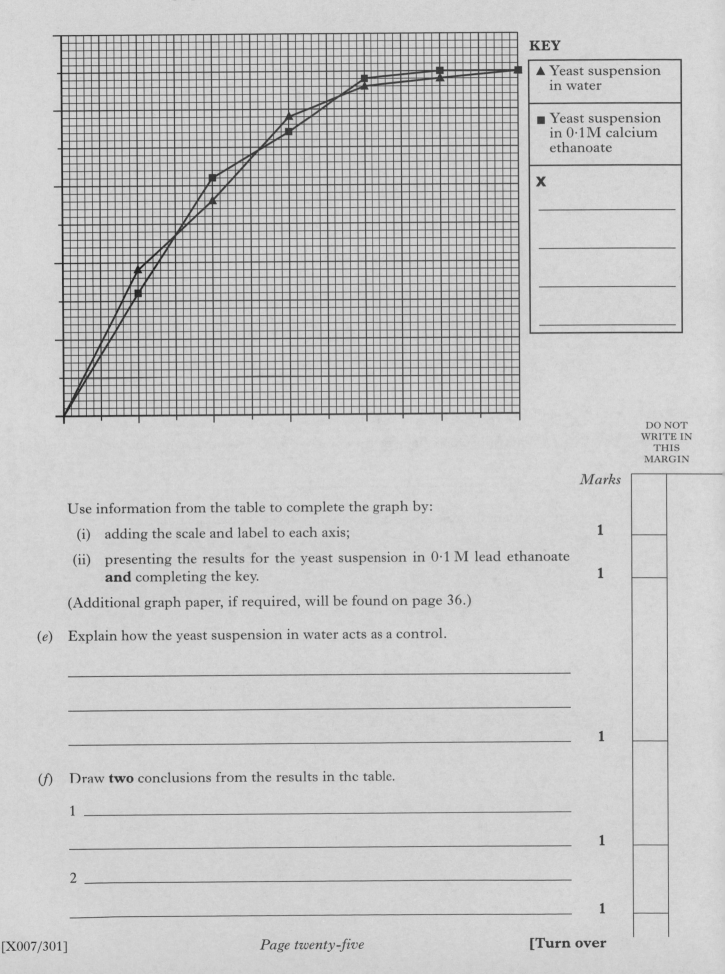

KEY

▲ Yeast suspension in water

■ Yeast suspension in 0·1 M calcium ethanoate

X

DO NOT WRITE IN THIS MARGIN

Marks

Use information from the table to complete the graph by:

(i) adding the scale and label to each axis; **1**

(ii) presenting the results for the yeast suspension in 0·1 M lead ethanoate **and** completing the key. **1**

(Additional graph paper, if required, will be found on page 36.)

(e) Explain how the yeast suspension in water acts as a control.

_____

_____

_____ **1**

(f) Draw **two** conclusions from the results in the table.

1 _____

_____ **1**

2 _____

_____ **1**

Marks

12. (*a*) The diagram shows a section through a barley grain.

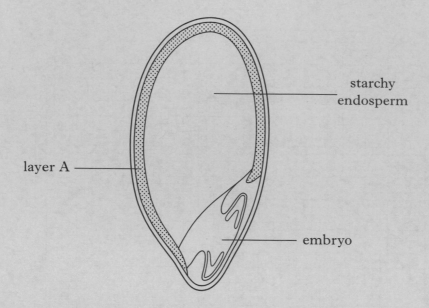

starchy
endosperm

layer A

embryo

Layer A produces α-amylase.

(i) Name layer A.

_____

1

(ii) What substance made by the embryo induces α-amylase production?

_____

1

(iii) Explain the role of α-amylase in the process of germination.

_____

_____

_____

2

(*b*) Give **one** practical application of a plant growth substance.

_____

1

Marks

13. The graph shows the results of an investigation into the relationship between environmental temperature and body temperature for a bobcat and a rattlesnake.

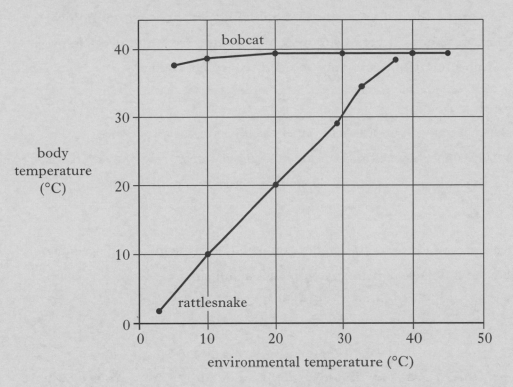

(a) Using information from the graph, **underline** one of the alternatives in each pair to make the sentence correct.

The rattlesnake is an $\left\{\begin{array}{l}\text{ectotherm}\\\text{endotherm}\end{array}\right\}$ because the results show that it $\left\{\begin{array}{l}\text{can}\\\text{cannot}\end{array}\right\}$ control its body temperature.

1

(b) Describe a rattlesnake behaviour pattern that is likely to raise its body temperature above the surrounding air temperature.

_____

_____ 1

(c) What evidence from the graph suggests that the bobcat has mechanisms to prevent overheating?

_____

_____ 1

(d) Explain why the bobcat's metabolic rate is greater at 10 °C than at 30 °C.

_____

_____

_____ 2

SECTION C

Marks

Both questions in this section should be attempted.

Note that each question contains a choice.

Questions 1 and 2 should be attempted on the blank pages which follow.

Supplementary sheets, if required, may be obtained from the invigilator.

Labelled diagrams may be used where appropriate.

1. Answer **either** A **or** B.

    A.   Give an account of populations under the following headings:

        (i)   the importance of monitoring wild populations;   **5**

        (ii)   the influence of density-dependent factors on population changes.   **5**

                                            (10)

    OR

    B.   Give an account of growth and development under the following headings:

        (i)   the influence of pituitary hormones in humans;   **4**

        (ii)   the effects of Indole Acetic Acid (IAA) in plants.   **6**

                                            (10)

In question 2, ONE mark is available for coherence and ONE mark is available for relevance.

2. Answer **either** A **or** B.

    A.   Give an account of the absorption of light energy by photosynthetic pigments and the light-dependent stage of photosynthesis.   **(10)**

    OR

    B.   Give an account of the structure of RNA and its role in protein synthesis.   **(10)**

[END OF QUESTION PAPER]

SPACE FOR ANSWERS

[Turn over

DO NOT
WRITE IN
THIS
MARGIN

SPACE FOR ANSWERS

SPACE FOR ANSWERS

SPACE FOR ANSWERS

DO NOT WRITE IN THIS MARGIN

[Turn over

SPACE FOR ANSWERS

SPACE FOR ANSWERS

DO NOT
WRITE IN
THIS
MARGIN

Page thirty-three

[Turn over

DO NOT
WRITE IN
THIS
MARGIN

SPACE FOR ANSWERS

DO NOT
WRITE IN
THIS
MARGIN

SPACE FOR ANSWERS

Page thirty-five

[Turn over

SPACE FOR ANSWERS

ADDITIONAL GRAPH PAPER FOR QUESTION 11(*d*)

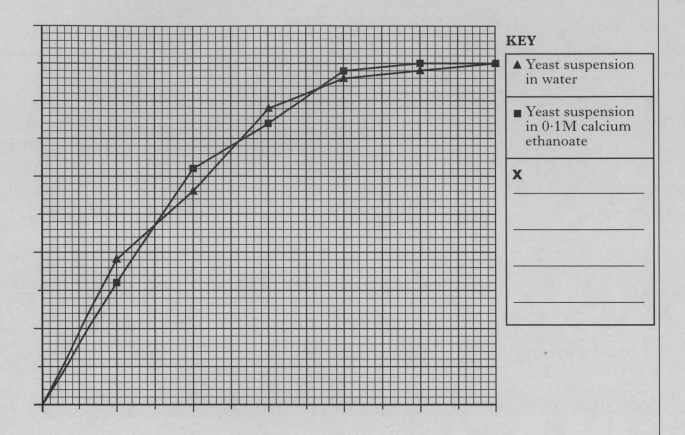

KEY

▲ Yeast suspension in water

■ Yeast suspension in 0·1M calcium ethanoate

X

[BLANK PAGE]

FOR OFFICIAL USE

Total for
Sections
B and C

X007/301

NATIONAL
QUALIFICATIONS
2006

TUESDAY, 23 MAY
1.00 PM – 3.30 PM

BIOLOGY
HIGHER

Fill in these boxes and read what is printed below.

Full name of centre

Town

Forename(s)

Surname

Date of birth
Day Month Year

Scottish candidate number

Number of seat

SECTION A—Questions 1–30 (30 marks)

Instructions for completion of Section A are given on page two.

For this section of the examination you must use an **HB pencil**.

SECTIONS B AND C (100 marks)

1 (a) All questions should be attempted.

  (b) It should be noted that in **Section C** questions 1 and 2 each contain a choice.

2 The questions may be answered in any order but all answers are to be written in the spaces provided in this answer book, **and must be written clearly and legibly in ink**.

3 Additional space for answers will be found at the end of the book. If further space is required, supplementary sheets may be obtained from the invigilator and should be inserted inside the **front** cover of this book.

4 The numbers of questions must be clearly inserted with any answers written in the additional space.

5 Rough work, if any should be necessary, should be written in this book and then scored through when the fair copy has been written. If further space is required a supplementary sheet for rough work may be obtained from the invigilator.

6 Before leaving the examination room you must give this book to the invigilator. If you do not, you may lose all the marks for this paper.

Read carefully

1   Check that the answer sheet provided is for **Biology Higher (Section A)**.

2   For this section of the examination you must use an **HB pencil**, and where necessary, an eraser.

3   Check that the answer sheet you have been given has **your name**, **date of birth**, **SCN** (Scottish Candidate Number) and **Centre Name** printed on it.

    Do not change any of these details.

4   If any of this information is wrong, tell the Invigilator immediately.

5   If this information is correct, **print** your name and seat number in the boxes provided.

6   The answer to each question is **either** A, B, C or D. Decide what your answer is, then, using your pencil, put a horizontal line in the space provided (see sample question below).

7   There is **only one correct** answer to each question.

8   Any rough working should be done on the question paper or the rough working sheet, **not** on your answer sheet.

9   At the end of the exam, put the **answer sheet for Section A inside the front cover of this answer book**.

Sample Question

The apparatus used to determine the energy stored in a foodstuff is a

A   calorimeter

B   respirometer

C   klinostat

D   gas burette.

The correct answer is **A**—calorimeter. The answer **A** has been clearly marked in **pencil** with a horizontal line (see below).

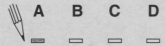

Changing an answer

If you decide to change your answer, carefully erase your first answer and using your pencil fill in the answer you want. The answer below has been changed to **D**.

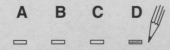

All questions in this section should be attempted.

Answers should be given on the separate answer sheet provided.

1. The diagram below represents a highly magnified section of a yeast cell.

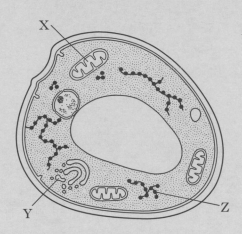

Which line of the table below correctly links each cell structure with its function?

	Aerobic respiration	Protein synthesis	Packaging materials for secretion
A	X	Y	Z
B	Y	Z	X
C	X	Z	Y
D	Z	X	Y

2. The action spectrum in photosynthesis is a measure of the ability of photosynthetic pigments to

   A absorb red and blue light

   B absorb light of different intensities

   C carry out photolysis

   D use light of different wavelengths for synthesis.

3. The diagram below shows the energy flow in an area of forest canopy during 1 year.

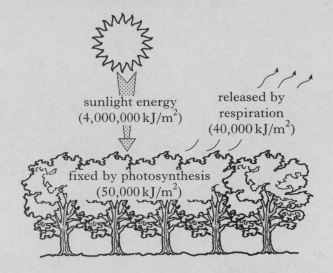

What percentage of available sunlight energy is fixed by the trees?

   A 0·25%

   B 1·00%

   C 1·25%

   D 2·25%

[Turn over

4. Photosynthetic pigments can be separated by means of chromatography as shown in the diagram below.

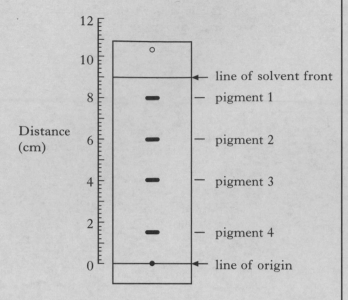

A pigment can be identified from its R_f value which can be calculated as follows:

$$R_f = \frac{\text{distance travelled by pigment from origin}}{\text{distance travelled by solvent from origin}}$$

Which line of the table correctly identifies the R_f values of pigments 2 and 3 on the above chromatogram?

	Pigment 2	Pigment 3
A	0·44	0·17
B	0·67	0·44
C	0·44	0·67
D	0·67	0·89

5. The diagram below shows a mitochondrion surrounded by cytoplasm.

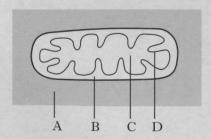

Where does glycolysis take place?

6. Which of the following statements refer to glycolysis?

1 Carbon dioxide released.

2 Occurs during aerobic respiration.

3 The end product is pyruvic acid.

4 The end product is lactic acid.

A 1 and 3

B 1 and 4

C 2 and 3

D 2 and 4

7. During anaerobic respiration in muscle fibres what is the fate of pyruvic acid?

A It is converted to lactic acid.

B It is broken down by the mitochondria.

C It is broken down to carbon dioxide and water.

D It is converted to citric acid.

8. Which of the following proteins has a fibrous structure?

A Insulin

B Pepsin

C Amylase

D Collagen

9. If ten percent of the bases in a molecule of DNA are adenine, what is the ratio of adenine to guanine in the same molecule?

A 1:1

B 1:2

C 1:3

D 1:4

10. In the life cycle of a bacterial virus which of the following sequences of events occurs?

A Lysis of the cell membrane, synthesis of viral DNA, replication of viral protein

B Lysis of cell membrane, synthesis of viral protein, replication of viral DNA

C Replication of viral DNA, synthesis of viral protein, lysis of cell membrane

D Synthesis of viral protein, replication of viral DNA, lysis of cell membrane

11. The table below shows some genotypes and phenotypes associated with forms of sickle-cell anaemia.

Phenotype	Genotype
unaffected	$Hb^A Hb^A$
sickle-cell trait	$Hb^A Hb^S$
acute sickle-cell anaemia	$Hb^S Hb^S$

A woman with sickle-cell trait and a man who is unaffected plan to have a child.

What are the chances that their child will have acute sickle-cell anaemia?

A   None

B   1 in 1

C   1 in 2

D   1 in 4

12. The following cross was carried out using two pure-breeding strains of the fruit fly, *Drosophila*.

P   straight wing   curly wing
              +   ×   +
           black body   grey body

F_1 All straight wing + black body

   The F_1 were allowed to interbreed

F_2 straight wing   curly wing
              +   ×   +
           black body   grey body

F_2 Phenotype
Ratio   3   :   1

In a dihybrid cross the typical F_2 ratio is 9 : 3 : 3 : 1.

An explanation of the result obtained in the above cross is that

A   crossing over has occurred between the genes

B   before isolation F_1 females had mated with their own type males

C   non-disjunction of chromosomes in the sex cells has taken place

D   these genes are linked.

13. The recombination frequency obtained in a genetic cross may be used as a source of information concerning the

A   genotypes of the recombinant offspring

B   diploid number of the species

C   fertility of the species

D   position of gene loci.

14. Red-green colour deficient vision is a sex-linked condition. John, who is affected, has the family tree shown below.

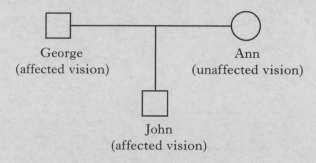

George
(affected vision)

Ann
(unaffected vision)

John
(affected vision)

If b is the mutant allele for the condition, which of the following could be the genotypes of George's parents and Ann's parents?

	George's parents		Ann's parents	
A	$X^B X^b$	$X^B Y$	$X^B X^B$	$X^B Y$
B	$X^B X^B$	$X^b Y$	$X^B X^B$	$X^B Y$
C	$X^B X^b$	$X^B Y$	$X^B X^b$	$X^B Y$
D	$X^B X^B$	$X^b Y$	$X^B X^B$	$X^b Y$

15. Klinefelter's syndrome is caused by the presence of an extra X chromosome in human males. Affected individuals are therefore XXY.

This syndrome is caused by

A   recombination

B   sex-linkage

C   crossing-over

D   non-disjunction.

[Turn over

16. The table refers to the mass of DNA in certain human body cells.

Cell type	Mass of DNA in cell ($\times 10^{-12}$ g)
liver	6·6
lung	6·6
R	3·3
S	0·0

Which of the following is the most likely identification of cell types R and S?

	R	S
A	ovum	mature red blood cell
B	mature red blood cell	sperm
C	nerve cell	mature red blood cell
D	kidney tubule cell	ovum

17. The following steps are involved in the process of genetic engineering.

1 Insertion of a plasmid into a bacterial host cell

2 Use of an enzyme to cut out a piece of chromosome containing a desired gene

3 Insertion of the desired gene into the bacterial plasmid

4 Use of an enzyme to open a bacterial plasmid

What is the correct sequence of these steps?

A   4   1   2   3

B   2   4   3   1

C   4   3   1   2

D   2   3   4   1

18. Osmoregulation in bony fish is achieved by a variety of strategies, depending on the nature of the environment.

Strategies

1 Large volume of dilute urine produced.

2 Small volume of concentrated urine produced.

3 Kidneys contain many large glomeruli.

4 Kidneys contain few small glomeruli.

Which of these strategies are employed by a freshwater bony fish?

A   1 and 4

B   2 and 4

C   1 and 3

D   2 and 3

19. The list below shows benefits which an animal species can obtain from certain types of social behaviour.

1 Aggression between individuals is controlled.

2 Subordinate animals are more likely to gain an adequate food supply.

3 Experienced leadership is guaranteed.

4 Energy used by individuals to obtain food is reduced.

Which statements refer to co-operative hunting?

A   1 and 2 only

B   1 and 3 only

C   2 and 4 only

D   3 and 4 only

20. The rates of carbon dioxide exchange by the leaves of two species of plants were measured at different light intensities.

The results are shown in the graph below.

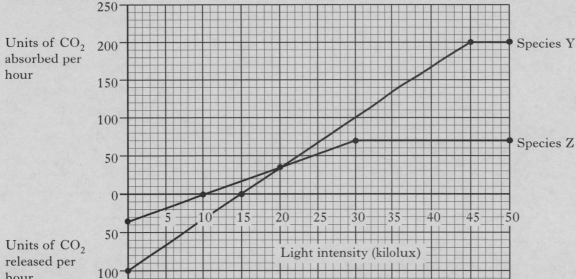

By how many kilolux is the compensation point for species Y greater than the compensation point for species Z?

A   5

B   10

C   15

D 130

21. A feature of phenylketonuria in humans is

A   the synthesis of excess phenylalanine

B   an inability to synthesise phenylalanine

C   the synthesis of excess tyrosine from phenylalanine

D   an inability to synthesise tyrosine from phenylalanine.

22. Plant ovary wall cells develop differently from plant phloem cells because of

A   random assortment in meiosis

B   genes being switched on and off during development

C   their having different numbers of chromosomes

D   their having different sets of genes.

[Turn over

23. An investigation was carried out into the germination of barley seeds.

The concentration of amylase and the rate of breakdown of starch was measured over 15 days.

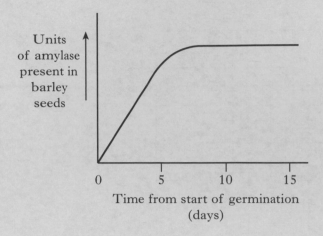

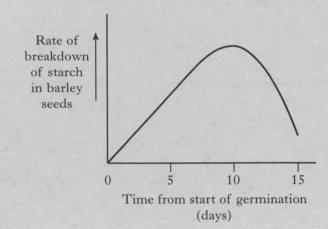

From the graphs it can be seen that after 10 days

A   the production of gibberellin has ceased

B   the rate of starch digestion decreases

C   the barley is synthesising its own starch

D   the amylase is becoming denatured.

24. The graph below contains information about fertiliser usage.

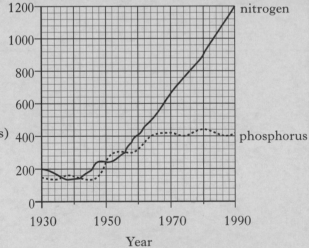

Which of the following statements about nitrogen usage between 1930 and 1990 is correct?

A   It increased steadily.

B   It increased by 500%.

C   It increased by 600%.

D   It always exceeded phosphorus usage.

25. Dietary deficiency of vitamin D causes rickets.

This effect is due to

A   poor uptake of phosphate into growing bones

B   low vitamin D content in the bones

C   poor calcium absorption from the intestine

D   loss of calcium from the bones.

26. Which of the following would result from an increased production of ADH?

A   The production of urine with a higher concentration of urea

B   Decrease in the permeability of the kidney collecting ducts

C   A decrease in the rate of glomerular filtration

D   An increase in the rate of production of urine

27. The graph below records the body temperature of a woman during an investigation in which her arm was immersed in water.

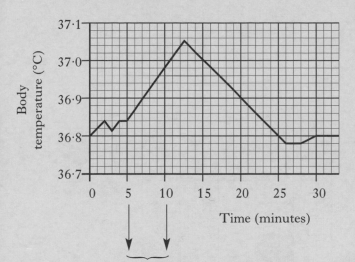

By how much did the temperature of her body vary during the 30 minutes of the investigation?

A 2·7 °C

B 0·27 °C

C 2·5 °C

D 0·25 °C

28. The graph below contains information about the birth rate and death rate in Mexico.

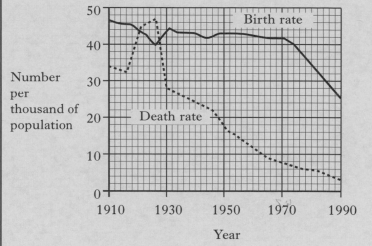

Which of the following conclusions can be drawn from the graph?

A At no time during the century has the population of Mexico decreased.

B The greatest increase in population occurred in 1970.

C The population was growing faster in 1910 than in 1990.

D Birth rate decreased between 1970 and 1990 due to the use of contraception.

[Turn over

29. The bar chart below shows the percentage loss in yield of four organically grown crops as a result of the effects of weeds, disease and insects.

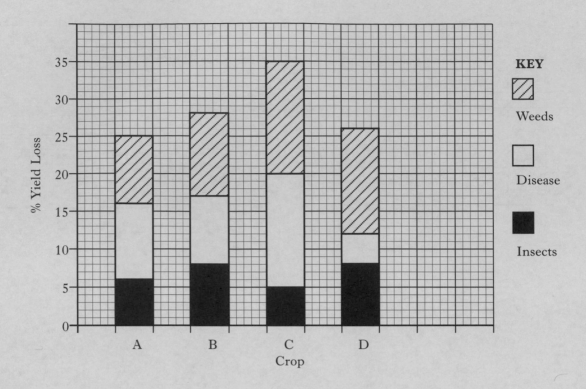

Predict which crop is most likely to show the greatest increase in yield if herbicides and insecticides were applied.

30. During succession in plant communities a number of changes take place in the ecosystem. Which line of the table correctly describes some of these changes?

	Species diversity	Biomass	Food web complexity
A	rises	rises	rises
B	rises	falls	rises
C	falls	rises	rises
D	rises	rises	falls

Candidates are reminded that the answer sheet MUST be returned INSIDE the front cover of this answer book.

[Turn over for Section B on *Page twelve*

SECTION B

All questions in this section should be attempted.

All answers must be written clearly and legibly in ink.

Marks

1. (a) The grid contains information about the plasma membrane and the cell wall.

A contains phospholipid	B fully permeable	C made of fibres
D contains cellulose	E selectively permeable	F made up of two layers

(i) Two of the boxes contain information about the **structure** of the plasma membrane.

Identify these **two** boxes.

Letters _____ and _____   1

(ii) One of the boxes contains information that relates to the role of the cell wall in the movement of water into a cell.

Identify this box.

Letter _____   1

(b) The table shows the percentage change in mass of apple tissue after immersion in sucrose solutions of different concentrations.

Concentration of sucrose solution (M)	Change in mass of apple tissue (%)
0·00	+22·0
0·10	+13·0
0·15	+8·5
0·20	+4·0
0·25	−0·5
0·30	−5·0
0·40	−14·0
0·50	−23·0

Marks

1. (*b*) (continued)

(i) Using values from the table, plot a line graph to show the percentage change in mass of the apple tissue against the concentration of sucrose solution.

Use appropriate scales to fill most of the grid.

(An additional grid, if required, may be found on page 40.)

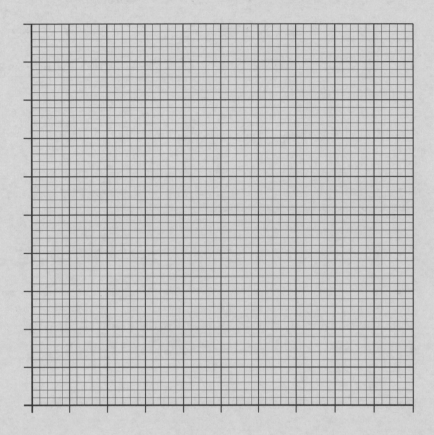

2

(ii) Complete the following sentences by underlining **one** of the alternatives in each pair.

The 0·1 M sucrose solution is $\left\{ \begin{array}{l} \text{hypotonic} \\ \text{hypertonic} \end{array} \right\}$ to the apple tissue.

Apple cells immersed in this solution may become $\left\{ \begin{array}{l} \text{plasmolysed} \\ \text{turgid} \end{array} \right\}$.

1

[Turn over

Marks

2. The diagram shows an outline of the stages in photosynthesis.

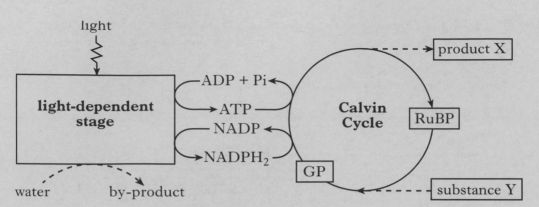

(*a*) (i) Name the by-product produced by the light-dependent stage.

_____ 1

(ii) Name product X and substance Y.

Product X _____

Substance Y _____ 1

(iii) State the number of carbon atoms in RuBP and GP.

RuBP _____ GP _____ 1

(*b*) (i) Describe the role of ATP in photosynthesis.

_____

_____

_____ 1

(ii) Explain why hydrogen from the light-dependent stage of photosynthesis is needed by the Calvin cycle.

_____

_____

_____ 1

2. **(continued)**

Marks

(c)   The graph shows the absorption spectra of three photosynthetic pigments.

Pigments P and Q were extracted from a hydrophyte with leaves that float on the water surface.

Pigment R was extracted from a species of photosynthetic algae that lives in the water below the hydrophyte.

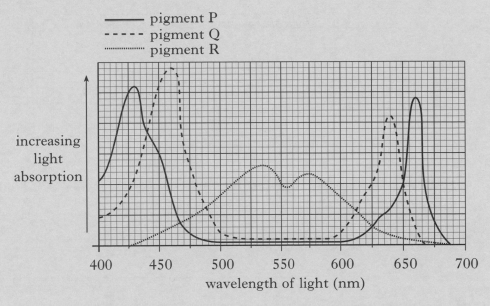

(i)   Tick (✓) **one** box to identify the wavelengths of light at which pigment P shows greatest absorption.

425–450 nm	450–475 nm	525–550 nm	625–650 nm	650–675 nm
☐	☐	☐	☐	☐

1

(ii)   Explain why it is an advantage to the hydrophyte to have more than one pigment.

_____

_____

_____

1

(iii)   Give **one** adaptation of hydrophyte leaves and state its effect.

Adaptation _____

Effect _____

_____

1

(iv)   From the information given, explain how the algae from which pigment R was extracted are adapted to photosynthesise in their environment.

_____

_____

_____

1

DO NOT WRITE IN THIS MARGIN

Marks

3. Experiments were carried out to investigate the hypothesis that the uptake of ions into mammalian cells takes place by active transport.

(*a*) The concentrations of potassium ions and chloride ions inside and outside a mammalian cell were measured.

The table shows the results obtained at an oxygen concentration of 4·0 units.

Ion	*Ion concentration inside cell* (mM)	*Ion concentration outside cell* (mM)
Potassium	140	5
Chloride	10	110

(i) Describe the information shown in the table that supports the original hypothesis.

____

____

____ **1**

(ii) From the table, calculate the simplest whole number ratio of potassium ions to chloride ions outside a mammalian cell.

Space for calculation

_______ potassium ions : _______ chloride ions **1**

Marks

3. **(continued)**

(*b*) The graph shows the effect of changing oxygen concentration on the concentration of potassium ions inside a mammalian cell.

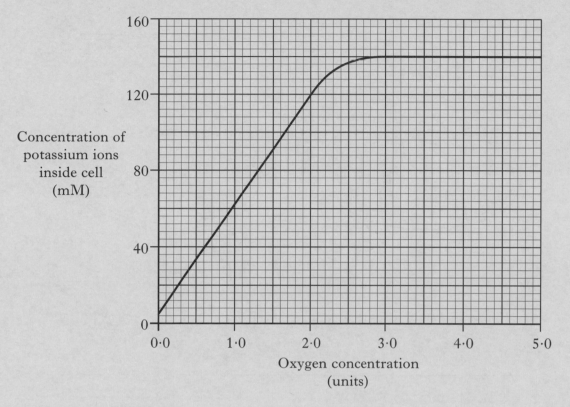

(i) Explain the shape of the graph between oxygen concentrations of 1·0 and 2·0 units.

_____

_____

_____

_____ 2

(ii) Suggest a reason why the graph levels off at oxygen concentrations above 3·0 units.

_____

_____ 1

[Turn over

DO NOT
WRITE IN
THIS
MARGIN

Marks

4. Yeast is a micro-organism capable of both aerobic and anaerobic respiration.

(*a*) Describe the role of oxygen in aerobic respiration.

_____

_____ **1**

(*b*) The diagrams represent a mitochondrion from a normal yeast cell and one from a mutant yeast cell.

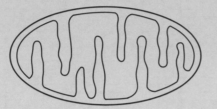

mitochondrion from   mitochondrion from
normal yeast cell   mutant yeast cell

(i) Name the structures that are absent from the mitochondrion of the mutant yeast cell.

_____ **1**

(ii) An experiment was carried out to investigate the effect of oxygen on the growth of normal and mutant yeast cells.

The method used in the experiment is outlined below.

- Three normal yeast cells were placed on agar growth medium containing glucose in a petri dish.

- Three mutant yeast cells were placed on the same agar growth medium containing glucose in a second dish.

- The dishes were then incubated at 30 °C for four days in aerobic conditions to allow the cells to multiply and produce colonies of yeast cells.

- The above three steps were repeated and the dishes were incubated this time in anaerobic conditions.

The sizes of the colonies produced are shown in the following diagrams.

Marks

4. **(b)** **(ii)** **(continued)**

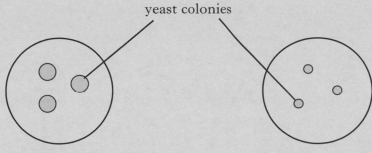

normal yeast
grown in aerobic conditions

mutant yeast
grown in aerobic conditions

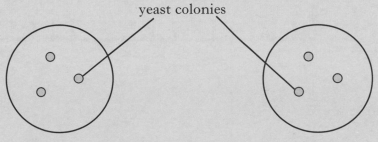

normal yeast
grown in anaerobic conditions

mutant yeast
grown in anaerobic conditions

1. Suggest a possible improvement to the experimental method, other than repeating the experiment, that would increase the reliability of the results.

_____

_____   1

2. Give an explanation for the difference in colony size observed for the normal yeast grown in aerobic and anaerobic conditions.

_____

_____

_____   2

3. Suggest why there was no difference in colony size when the mutant cells were grown in aerobic and anaerobic conditions.

_____

_____   1

[Turn over

Marks

5. The diagram shows translation of part of a mRNA molecule during the synthesis of a protein.

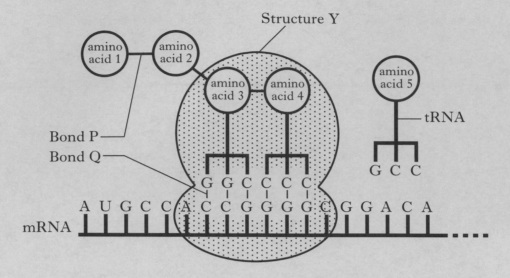

(a) Name structure Y.

_____   1

(b) Name the types of bond shown at P and Q.

Bond P _____

Bond Q _____   1

(c) Give the anticodon for the tRNA for amino acid 1.

_____   1

(d) Describe **two** functions of tRNA in protein synthesis.

1 _____

_____   1

2 _____

_____   1

(e) Genetic information for protein synthesis is in the form of a triplet code. Explain what is meant by this statement.

_____

_____

_____   1

Marks

6. Patients who have had tissue transplants may be treated with a drug that suppresses the immune system.

The table shows the number of lymphocytes in the blood of a patient before and after treatment.

Number of lymphocytes before treatment (cells per mm^3)	Number of lymphocytes after treatment (cells per mm^3)
7500	3000

(*a*) Calculate the percentage decrease in the number of lymphocytes following treatment with the suppressor drug.

Space for calculation

————————— % **1**

(*b*) Explain why there is a risk of rejection when tissues are transplanted.

_____

_____

_____

_____ **2**

(*c*) Some suppressor drugs act by binding to DNA molecules in such a way that the separation of the two DNA strands is prevented.

Predict **one** possible consequence of the use of this type of suppressor drug on the normal functions of DNA.

_____

_____

_____ **1**

[Turn over

Marks

7. (*a*) The letters A – E represent five statements about meiosis.

Letter	Statement
A	haploid gametes are produced
B	chiasmata are formed
C	chromatids separate
D	gamete mother cell is present
E	homologous chromosomes form pairs

(i) Use **all** the letters from the list to complete the table to show the statements that are connected with the first and the second meiotic divisions.

First meiotic division	*Second meiotic division*

2

(ii) Independent assortment of chromosomes during meiosis is a source of genetic variation.

Describe the behaviour of chromosomes during the first meiotic division stage that results in independent assortment.

_____

_____

_____   1

7. (continued)

Marks

(b) Duchenne muscular dystrophy (DMD) is a recessive, sex-linked condition in humans that affects muscle function.

The diagram shows an X-chromosome from an unaffected individual and one from an individual with DMD.

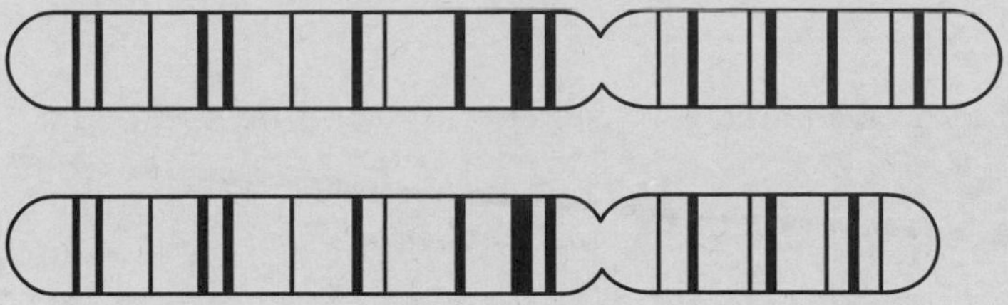

X-chromosome from unaffected individual

X-chromosome from individual with DMD

(i) Using information from the diagram, name the type of chromosome mutation responsible for DMD.

_____   1

(ii) **On the diagram** of the chromosome from the **unaffected individual**, put a **cross (X)** on the likely location of the gene involved in DMD.   1

(iii) Males are more likely to be affected by DMD than females.

Explain why.

_____

_____

_____

_____   1

(iv) A person with DMD has an altered phenotype compared with an unaffected individual.

Explain how an inherited chromosome mutation such as DMD may result in an altered phenotype.

_____

_____

_____

_____   1

[Turn over

Marks

8. Three species of small bird, the Blue Tit, the Great Tit and the Marsh Tit, forage for caterpillars in oak woods in early summer.

   Investigators observed each bird species for ten hours. They recorded the percentage of time each species spent foraging in different parts of the trees. The results are shown in the **Bar Chart**.

   Table 1 shows the percentage of the birds' diet that came from different caterpillar size ranges.

   Table 2 shows the beak size index calculated for each species using the following formula.

   beak size index = average beak length (mm) × average beak depth (mm)

Bar Chart

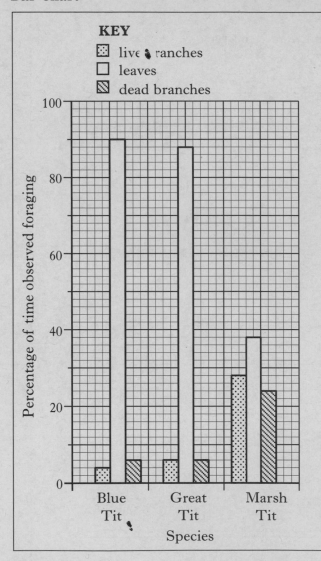

Table 1

caterpillar size range (mm)	% of diet from each size range		
	Blue Tit	Great Tit	Marsh Tit
1–2	63·6	18·2	18·1
3–4	24·3	20·4	60·6
5–6	10·0	27·2	12·1
7–8	2·1	34·2	9·2

Table 2

species	beak size index (mm^2)
Blue Tit	40·3
Great Tit	67·6
Marsh Tit	44·2

(a) Calculate the number of minutes the Blue Tits were observed foraging on live branches.

   Space for calculation

_____ mins   1

8. **(continued)**

Marks

(b) What evidence from the **Bar Chart** suggests that Marsh Tits forage on tree parts other than those shown?

_____

_____   1

(c) Calculate the average percentage of the birds' diets in the 1–2 mm caterpillar size range.

Space for calculation

_____ %   1

(d) Describe the relationship between beak size index and caterpillar size range eaten.

_____

_____   1

(e) The average beak length of Great Tits is 13 mm.

Calculate their average beak depth.

Space for calculation

_____ mm   1

(f) From the information given, describe all of the ways by which interspecific competition for caterpillars is reduced between the following pairs of bird species.

(i) Blue Tits and Great Tits _____

_____

_____   1

(ii) Blue Tits and Marsh Tits _____

_____

_____   1

(g) Each bird species must forage economically.

Explain what is meant by this statement in terms of energy gain and loss.

_____

_____

_____   1

Marks

9. The ancestor species of the modern tomato produces small fruits but can grow in soils with low nitrogen levels.

The modern species of tomato has been selectively bred to produce large fruit but requires soil rich in nitrogen to grow well.

(*a*) (i) Give **one** reason why nitrogen is important for plant growth.

_____

_____   1

(ii) Give **one** symptom of nitrogen deficiency in plants.

_____   1

(*b*) The diagram represents steps in a technique used by tomato breeders to combine characteristics of these two species.

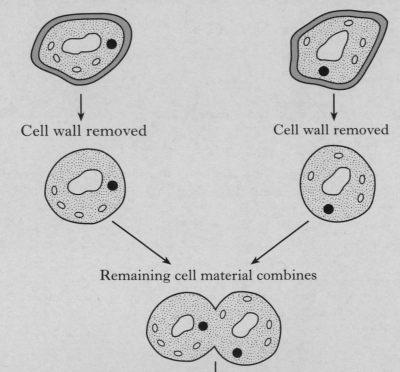

Leaf cell from ancestor plant   Leaf cell from selectively bred plant

Cell wall removed   Cell wall removed

Remaining cell material combines

New species of tomato plant

Marks

9. (*b*) **(continued)**

    (i) Give the name of the technique shown in the diagram.

_____   **1**

    (ii) Name the enzyme that is used to remove the cell walls from the leaf cells.

_____   **1**

    (iii) Explain why cultivation of the new species of tomato could lead to a reduction in the use of nitrogen fertiliser.

_____

_____

_____   **1**

[Turn over

Marks

10. An investigation was carried out to compare the growth of *Escherichia coli* (*E. coli*) bacteria in different nutrient solutions.

E. coli were grown in a glucose solution for 24 hours. Two 50 cm³ samples were transferred to two identical containers each with different sterile nutrient solutions as shown in the table.

Container	Nutrient solution
X	0·5 mM glucose
Y	0·5 mM lactose

One container is shown in the diagram.

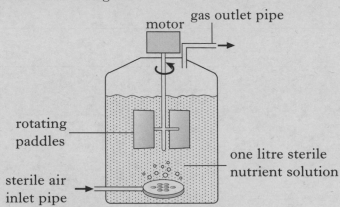

The pH and temperature were kept constant.

Every 30 minutes, a 2 cm³ sample was taken from each container.

An instrument was used to measure the number of bacteria present. The higher the instrument reading, the more bacteria.

(*a*) (i) Suggest a reason for having rotating paddles.

_____

_____   1

(ii) Explain why a gas outlet pipe is needed in the apparatus.

_____

_____

_____   1

(*b*) Identify **two** variables not already mentioned that would have to be controlled in both containers to make the procedure valid.

1 _____   1

2 _____   1

Marks

10. (continued)

(c)   The results of the investigation are shown in the graph.

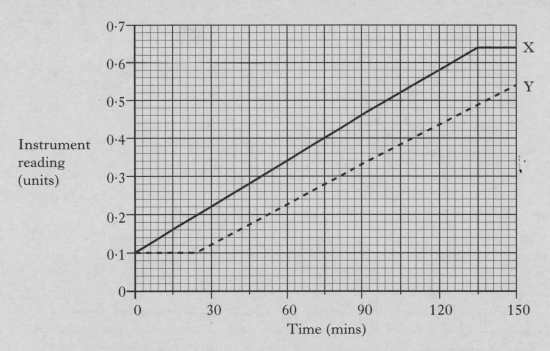

E. coli only produces the enzyme to metabolise lactose when there is no glucose available to the cells.

(i)   Use information from the graph to state the time taken for the bacterial cells to produce the enzyme needed to metabolise lactose.

Justify your answer.

Time _____ minutes

Justification _____

_____

_____   1

(ii)   Explain how lactose acts as an inducer of this enzyme.

_____

_____

_____   2

(iii)   Predict the instrument reading at 150 minutes if a third container had been used with nutrient solution containing 0·25 mM glucose.

_____ units   1

DO NOT WRITE IN THIS MARGIN

Marks

11. In seed pods of garden pea plants, smooth shape (T) is dominant to constricted shape (t) and green (G) is dominant to yellow (g).

The genes are not linked.

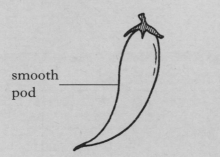
smooth pod

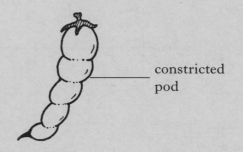
constricted pod

(a) In an investigation, a pea plant heterozygous for both smooth and green pods was crossed with a pea plant with constricted, yellow pods.

Complete the table to give the genotypes of the parent plants and **all** of their possible gametes.

Phenotype of parent	*Genotype of parent*	*Genotype(s) of gamete(s)*
Smooth green pod		
Constricted yellow pod		

1

1

(b) In a second investigation, two pea plants heterozygous for both seed shape and seed colour were crossed. This produced 112 offspring.

(i) Calculate the **expected** number of offspring that would have yellow pods.

Space for calculation

_____ 1

The actual phenotypes obtained did not occur in the expected numbers.

(ii) Suggest **two** reasons why the **actual** numbers observed may differ from the **expected** numbers in a dihybrid cross.

1 _____

_____ 1

2 _____

_____ 1

Marks

12. A potato tuber is a swollen underground stem.

Two similar tubers, each with one apical bud and four lateral buds were treated as shown in the diagram. Their appearance after being kept in the dark for three weeks is also shown.

Apical buds produce the plant growth substance IAA.

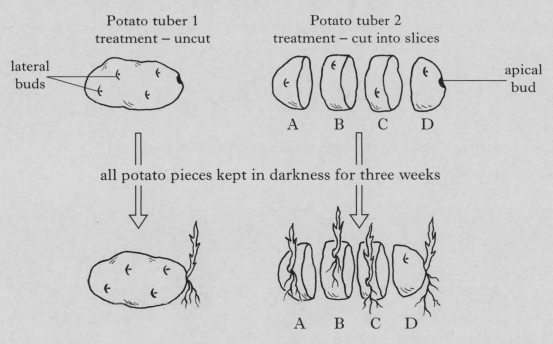

(a) Name the effect of IAA that is shown by potato tuber 1 after three weeks.

_____ 1

(b) Explain why the lateral buds on slices A, B and C produced shoots.

_____

_____

_____ 1

(c) Potato tuber 1 was then exposed to light from one direction for a further three days. The shoot showed phototropism.

Explain the role of IAA in phototropism.

_____

_____

_____ 2

(d) Potatoes are long day plants. In terms of changing photoperiod, describe what is meant by "long day plant".

_____

_____ 1

Marks

13. The table shows the mass of salmon caught in coastal waters around Scotland.

Year	Scottish catch (tonnes)
1973	1300
1980	1200
1988	900
1998	200

(a) By how many times was the 1973 catch greater than the 1998 catch?

Space for calculation

_____ times   **1**

(b) Scientists monitoring the salmon have suggested the following four possible factors for the decline in numbers.

- Predation by seals

- Food shortage

- Rising sea temperature

- Infection by sea louse parasites

Underline the factor(s) that would have had a density-independent effect on the salmon population.   **1**

(c) Populations of North Atlantic salmon are monitored because they are a food species.

Give **one** other reason for monitoring animal populations.

_____

_____   **1**

Marks

14. The table shows how two structures in mammalian skin respond to a drop in the surrounding air temperature from 20 °C to 5 °C.

Structure	Air temperature	
	20 °C	5 °C
hair erector muscles	relaxed	contracted
blood vessels	dilated	constricted

(a) (i) Name the temperature monitoring centre in the body of a mammal.

_____   1

(ii) State how messages are sent from the temperature monitoring centre to the skin.

_____   1

(b) Explain the advantage to the organism of constriction of skin blood vessels when the air temperature drops from 20 °C to 5 °C.

_____

_____

_____

_____   1

(c) Give the term that describes an animal that obtains most of its body heat from its own metabolism.

_____   1

[Turn over for Section C on *Page thirty-four*

Marks

SECTION C

Both questions in this section should be attempted.

Note that each question contains a choice.

Questions 1 and 2 should be attempted on the blank pages which follow.

Supplementary sheets, if required, may be obtained from the invigilator.

All answers must be written clearly and legibly in ink.

Labelled diagrams may be used where appropriate.

1. Answer **either** A **or** B.

   A. Give an account of transpiration under the following headings:

      (i)   the effect of environmental factors on transpiration rate;   **5**

      (ii) adaptations of xerophyte plants that reduce the transpiration rate.   **5**

               (10)

   OR

   B. Give an account of how animals and plants cope with dangers under the following headings:

      (i)   behavioural defence mechanisms in animals;   **5**

      (ii) cellular and structural defence mechanisms in plants.   **5**

               (10)

In question 2, ONE mark is available for coherence and ONE mark is available for relevance.

2. Answer **either** A **or** B.

   A. Give an account of the principle of negative feedback with reference to the maintenance of blood sugar levels.   **(10)**

   OR

   B. Give an account of the role of the pituitary gland in controlling normal growth and development and describe the effects of named drugs on fetal development.   **(10)**

[END OF QUESTION PAPER]

SPACE FOR ANSWERS

DO NOT
WRITE IN
THIS
MARGIN

[Turn over

SPACE FOR ANSWERS

SPACE FOR ANSWERS

[Turn over

DO NOT
WRITE IN
THIS
MARGIN

SPACE FOR ANSWERS

SPACE FOR ANSWERS

[Turn over

SPACE FOR ANSWERS

ADDITIONAL GRAPH PAPER FOR QUESTION 1(*b*)(i)

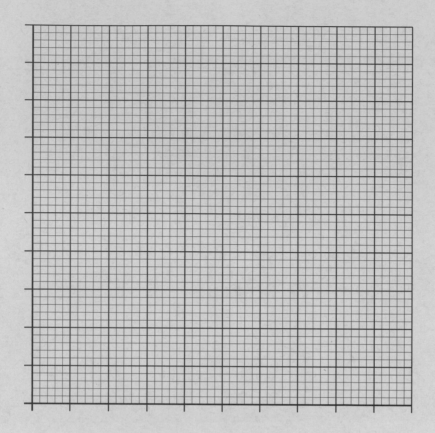

[BLANK PAGE]

FOR OFFICIAL USE

Total for
Sections
B and C

X007/301

NATIONAL
QUALIFICATIONS
2007

MONDAY, 21 MAY
1.00 PM – 3.30 PM

BIOLOGY
HIGHER

Fill in these boxes and read what is printed below.

Full name of centre

Town

Forename(s)

Surname

Date of birth
Day Month Year

Scottish candidate number

Number of seat

SECTION A—Questions 1–30 (30 marks)

Instructions for completion of Section A are given on page two.

For this section of the examination you must use an **HB pencil**.

SECTIONS B AND C (100 marks)

1   (a)   All questions should be attempted.

(b)   It should be noted that in **Section C** questions 1 and 2 each contain a choice.

2   The questions may be answered in any order but all answers are to be written in the spaces provided in this answer book, **and must be written clearly and legibly in ink**.

3   Additional space for answers will be found at the end of the book. If further space is required, supplementary sheets may be obtained from the invigilator and should be inserted inside the **front** cover of this book.

4   The numbers of questions must be clearly inserted with any answers written in the additional space.

5   Rough work, if any should be necessary, should be written in this book and then scored through when the fair copy has been written. If further space is required a supplementary sheet for rough work may be obtained from the invigilator.

6   Before leaving the examination room you must give this book to the invigilator. If you do not, you may lose all the marks for this paper.

Scottish
Qualifications
Authority

Read carefully

1 Check that the answer sheet provided is for **Biology Higher (Section A)**.

2 For this section of the examination you must use an **HB pencil**, and where necessary, an eraser.

3 Check that the answer sheet you have been given has **your name**, **date of birth**, **SCN** (Scottish Candidate Number) and **Centre Name** printed on it.

   Do not change any of these details.

4 If any of this information is wrong, tell the Invigilator immediately.

5 If this information is correct, **print** your name and seat number in the boxes provided.

6 The answer to each question is **either** A, B, C or D. Decide what your answer is, then, using your pencil, put a horizontal line in the space provided (see sample question below).

7 There is **only one correct** answer to each question.

8 Any rough working should be done on the question paper or the rough working sheet, **not** on your answer sheet.

9 At the end of the exam, put the **answer sheet for Section A inside the front cover of this answer book**.

Sample Question

The apparatus used to determine the energy stored in a foodstuff is a

A calorimeter

B respirometer

C klinostat

D gas burette.

The correct answer is **A**—calorimeter. The answer **A** has been clearly marked in **pencil** with a horizontal line (see below).

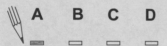

Changing an answer

If you decide to change your answer, carefully erase your first answer and using your pencil fill in the answer you want. The answer below has been changed to **D**.

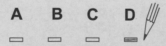

SECTION A

All questions in this section should be attempted.

Answers should be given on the separate answer sheet provided.

1. Which line in the table identifies correctly the two cell structures shown in the diagram?

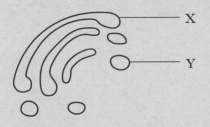

	X	Y
A	Golgi body	Vesicle
B	Golgi body	Ribosome
C	Endoplasmic reticulum	Vesicle
D	Endoplasmic reticulum	Ribosome

2. The phospholipid molecules in a cell membrane allow the

    A free passage of glucose molecules

    B self-recognition of cells

    C active transport of ions

    D membrane to be fluid.

3. Red blood cells have a solute concentration of around 0·9%.

    Which of the following statements correctly describes the fate of these cells when immersed in a 1% salt solution?

    A The cells will burst.

    B The cells will shrink.

    C The cells will expand but not burst.

    D The cells will remain unaffected.

4. Which graph best illustrates the effect of increasing temperature on the rate of active uptake of ions by roots?

A
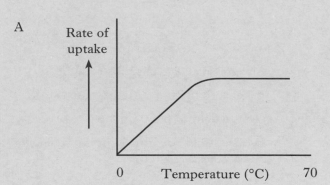

B
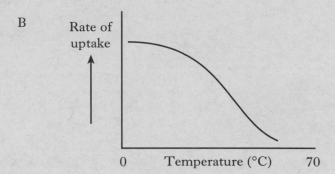

C
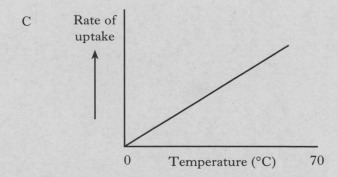

D
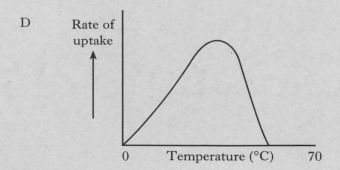

5. Which substances must be provided by host cells for the synthesis of viruses?

A Proteins and nucleotides

B Amino acids and DNA

C Proteins and DNA

D Amino acids and nucleotides

6. The action spectrum of photosynthesis is a measure of the ability of plants to

A absorb all wavelengths of light

B absorb light of different intensities

C use light to build up foods

D use light of different wavelengths for synthesis.

7. The diagram shows DNA during replication. Base H represents thymine and base M represents guanine. Which letters represent the base cytosine?

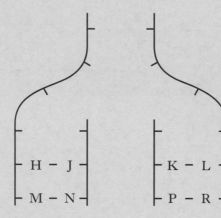

A J and K

B J and L

C N and P

D N and R

8. The graph below shows the effect of light intensity on the rate of photosynthesis at different temperatures.

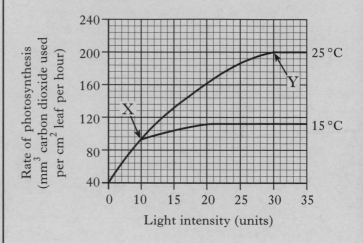

Which of the following conclusions can be made from the above data?

A Only at light intensities greater than 20 units does temperature affect the rate of photosynthesis.

B At point Y, the rate of photosynthesis is limited by the light intensity.

C Temperature has little effect on the rate of photosynthesis at low light intensities.

D At point X, temperature limits the rate of photosynthesis.

9. The cell structures shown below have been magnified ten thousand times.

Mitochondrion   Chloroplast

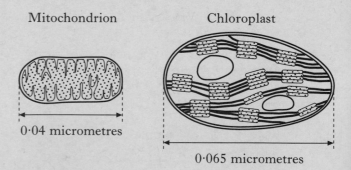

0·04 micrometres   0·065 micrometres

Expressed as a simple whole number ratio, the length of the mitochondrion compared to that of the chloroplast is

A   8 : 13

B   13 : 8

C   40 : 65

D   65 : 40.

10. A section of a DNA molecule contains 300 bases. Of these bases, 90 are adenine. How many cytosine bases would this section of DNA contain?

A 60

B 90

C 120

D 180

11. What information can be derived from the recombination frequencies of linked genes?

A The mutation rate of the genes

B The order and location of genes on a chromosome

C Whether genes are recessive or dominant

D The genotype for a particular characteristic

12. The diagram shows a family tree for a family with a history of red-green colour deficiency.

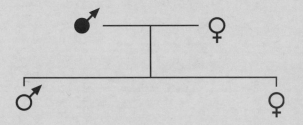

 Unaffected male

 Affected male

 Unaffected female

 Affected female

The allele for red-green colour deficiency is sex-linked.

Which of the following statements is true?

A Only the son is a carrier.

B Only the daughter is a carrier.

C Both son and daughter are carriers.

D Neither son nor daughter is a carrier.

13. Apple crop yields have been increased by plant breeders selecting for

A disease resistance

B flavour

C resistance to bruising

D sugar content.

14. Human insulin can be produced by the bacterium *E. coli* using the following steps.

1 Culture large quantities of *E. coli* in vats of nutrients.

2 Insert human insulin gene into *E. coli* plasmid DNA.

3 Cut insulin gene from human chromosome using enzymes.

4 Extract insulin from culture vats.

The correct order for these steps is

A 3, 2, 1, 4

B 3, 1, 2, 4

C 1, 4, 3, 2

D 1, 2, 3, 4.

15. In a desert mammal, which of the following is a physiological adaptation which helps to conserve water?

A Nocturnal foraging

B Breathing humid air in a burrow

C Having few sweat glands

D Remaining underground by day

16. The following factors affect the transpiration rate in a plant.

1 increasing wind speed

2 decreasing humidity

3 rising air pressure

4 falling temperature

Which two of these factors would cause an increase in transpiration rate?

A 1 and 2

B 1 and 3

C 2 and 4

D 3 and 4

[Turn over

17. When first exposed to a harmless stimulus, a group of animals responded by showing avoidance behaviour. When the stimulus was repeated the animals became habituated to it.

What change in response would have shown that habituation was taking place?

A An increase in the length of the response

B A decrease in the time taken to respond

C An increase in response to other stimuli

D A decrease in the percentage of animals responding

18. In tomato plants, the allele for curled leaves is dominant over the allele for straight leaves. The allele for hairy stems is dominant over the allele for hairless stems. The genes for curliness and hairiness are located on different chromosomes.

If plants heterozygous for both characteristics were crossed, what ratio of phenotypes would be expected in the offspring?

A All curly and hairy

B 3 curly and hairy: 1 straight and hairless

C 9 curly and hairy: 3 curly and hairless:
3 straight and hairy: 1 straight and hairless

D 1 curly and hairy: 1 curly and hairless:
1 straight and hairy: 1 straight and hairless

19. The bar graph below shows changes in the DNA content per cell during stages of meiosis.

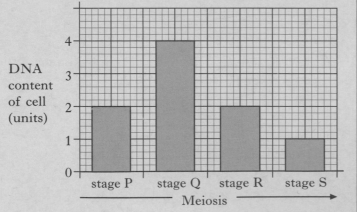

When do the homologous pairs of chromosomes separate?

A Before the start of stage P

B Between stages P and Q

C Between stages Q and R

D Between stages R and S

20. The genes for two different characteristics are located on separate chromosomes.

In a cross between individuals with the genotypes AaBb and aabb, what is the chance of any one of the offspring having the genotype aabb?

A 0

B 1 in 2

C 1 in 4

D 1 in 8

21. Which line in the table identifies correctly the main source of body heat and the method of controlling body temperature in an endotherm?

	Main source of body heat	*Principal method of controlling body temperature*
A	Respiration	Behavioural
B	Respiration	Physiological
C	Absorbed from environment	Behavioural
D	Absorbed from environment	Physiological

22. The following four statements relate to meristems.

1 Some provide cells for increase in diameter in stems

2 Some produce growth substances

3 They are found in all growing organisms

4 Their cells undergo division by meiosis

Which of the above statements are true?

A 1 and 2 only

B 1 and 3 only

C 2 and 3 only

D 2 and 4 only

23. The diagram below shows a transverse section of a woody stem.

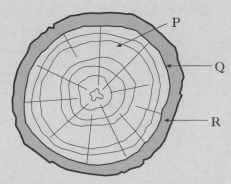

Which line of the table identifies correctly the tissues P, Q and R?

	P	Q	R
A	cambium	xylem	phloem
B	phloem	xylem	cambium
C	xylem	phloem	cambium
D	xylem	cambium	phloem

24. The graph below shows changes which occur in the masses of protein, fat and carbohydrate in a girl's body during seven weeks of starvation.

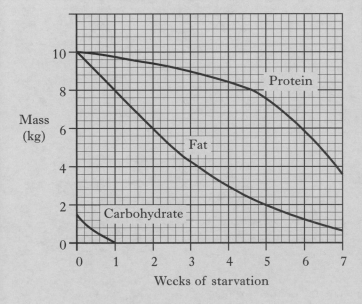

The girl weighs 60 kg at the start. Predict her weight after two weeks without food.

A 43 kg

B 50 kg

C 54 kg

D 57 kg

25. Which of the following is triggered by the hypothalamus in response to an increase in the temperature of the body?

A Contraction of the hair erector muscles and vasodilation of the skin capillaries

B Relaxation of the hair erector muscles and vasodilation of the skin capillaries

C Contraction of the hair erector muscles and vasoconstriction of the skin capillaries

D Relaxation of the hair erector muscles and vasoconstriction of the skin capillaries

26. Plants require macro-elements for the synthesis of various compounds. Identify which macro-elements are required for synthesis of the compounds shown in the table below.

	Chlorophyll	Protein	ATP
A	phosphorus	magnesium	nitrogen
B	phosphorus	nitrogen	magnesium
C	magnesium	nitrogen	phosphorus
D	magnesium	phosphorus	nitrogen

27. The following are events occurring during germination in barley.

1 The embryo produces gibberellic acid (GA)

2 α-amylase is produced

3 Gibberellic acid (GA) passes to the aleurone layer

4 α-amylase converts starch to maltose

5 Maltose is used by the embryo

Which of the following indicates the correct sequence of events?

A 1 2 4 5 3

B 1 3 2 4 5

C 3 2 1 4 5

D 5 1 3 2 4

[Turn over

28. The table shows the masses of various substances in the glomerular filtrate and in the urine over a period of 24 hours.

Which of the substances has the smallest percentage of reabsorption from the glomerular filtrate?

	Substance	Mass in glomerular filtrate (g)	Mass in urine (g)
A	Sodium	600·0	6·0
B	Potassium	35·0	2·0
C	Uric acid	8·5	0·8
D	Calcium	5·0	0·2

29. An investigation was carried out into the effect of indole acetic acid (IAA) concentration on the shoot growth of two species of plant. The graph below shows a summary of the results.

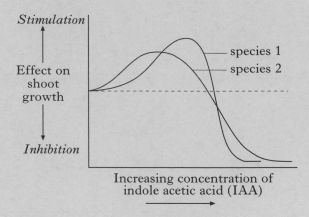

Which one of the following conclusions is justified?

A   Species 1 shows its maximum stimulation at a lower IAA concentration than species 2.

B   Species 2 is more inhibited by the highest concentrations of IAA than species 1.

C   Species 2 is stimulated over a greater range of IAA concentrations than species 1.

D   Species 1 is stimulated by some IAA concentrations which inhibit species 2.

30. An enzyme and its substrate were incubated with various concentrations of either copper or magnesium salts.

The time taken for the complete breakdown of the substrate was measured.

The results are given in the table.

Salt Concentration (M)	Time needed to break down substrate (s)	
	Copper salts	Magnesium salts
0	39	39
1×10^{-8}	42	21
1×10^{-6}	380	49
1×10^{-4}	1480	286

increasing concentration ↓

From the data, it may be deduced that

A   high concentrations of copper salts promote the activity of the enzyme

B   high concentrations of copper salts inhibit the activity of the enzyme

C   low concentrations of magnesium salts inhibit the activity of the enzyme

D   high concentrations of magnesium salts promote the activity of the enzyme.

Candidates are reminded that the answer sheet MUST be returned INSIDE the front cover of this answer book.

[Turn over for Section B on *Page ten*

Marks

SECTION B

All questions in this section should be attempted.

All answers must be written clearly and legibly in ink.

1. (*a*) The diagram contains information about light striking a leaf.

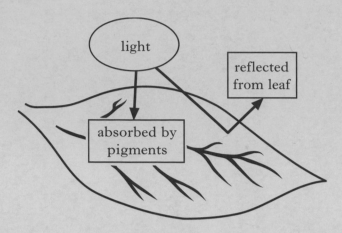

   (i) Apart from being absorbed or reflected, what can happen to light which strikes a leaf?

_____   1

   (ii) Pigments that absorb light are found within leaf cells.

      State the exact location of these pigments.

_____   1

(*b*) The diagram below shows part of the light dependent stage of photosynthesis.

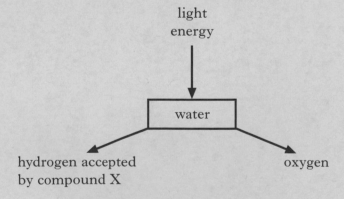

   (i) Name this part of the light dependent stage.

_____   1

   (ii) Name compound X.

_____   1

Marks

1. **(continued)**

(*c*) The following sentences describe events in the carbon fixation stage of photosynthesis.

Underline one alternative in each pair to make the sentences correct.

The $\left\{\begin{array}{c}\text{three}\\\text{five}\end{array}\right\}$ carbon compound ribulose bisphosphate (RuBP) accepts

$\left\{\begin{array}{c}\text{carbon dioxide}\\\text{hydrogen}\end{array}\right\}$.

$\left\{\begin{array}{c}\text{Carbon dioxide}\\\text{Hydrogen}\end{array}\right\}$ is accepted by the $\left\{\begin{array}{c}\text{three}\\\text{five}\end{array}\right\}$ carbon compound

glycerate phosphate (GP).

2

[Turn over

Marks

2. The diagram shows apparatus set up to investigate the rate of respiration in an earthworm. After 10 minutes at 20 °C the level of liquid in the capillary tube had changed as shown.

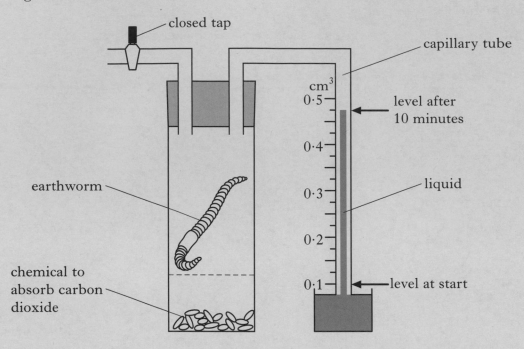

(*a*)   (i)   What volume of oxygen is used by the earthworm during the 10 minute period?

_____ cm^3 **1**

       (ii)   Describe a suitable control for this experiment.

_____

_____   **1**

(*b*)   In a second experiment, a worm of 5 grams used 0·5 cm^3 of oxygen in 10 minutes.

Calculate its rate of respiration in cm^3 per minute per gram of worm.

Space for calculation

_____ cm^3 per minute per gram of worm   **1**

[Turn over for Question 3 on *Page fourteen*

Marks

3. (*a*) Samples of carrot tissue were immersed in a hypotonic solution at two different temperatures for 5 hours. The mass of the tissue samples was measured every hour and the percentage change in mass calculated.

The results are shown on the graph.

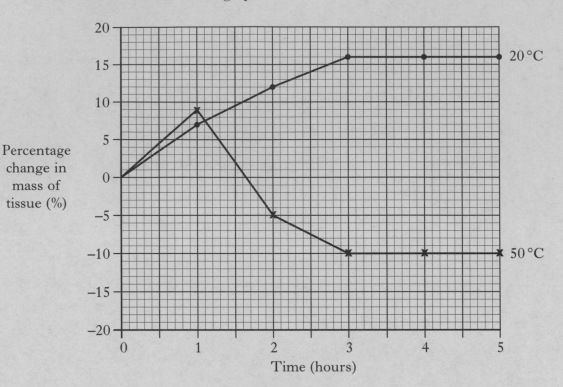

(i) Explain the results obtained at 20 °C from 0 to 3 hours and from 3 hours to 5 hours.

0 to 3 hours _____

_____ **1**

3 to 5 hours _____

_____ **1**

(ii) Explain the change in mass of the carrot tissue between 1 and 3 hours at 50 °C.

_____

_____

_____ **2**

Marks

3.   (continued)

(*b*)   The chart shows the concentration of ions within a unicellular organism and in the sea water surrounding it.

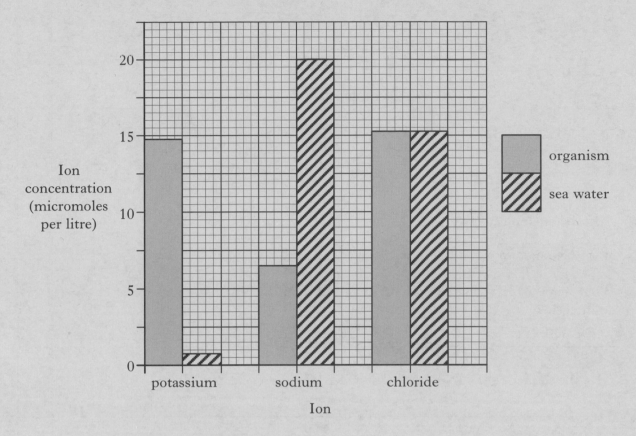

(i)   From the information given, identify the ion which appears to move between the organism and the sea water by diffusion.

Justify your choice.

Ion _____

Justification _____

_____   1

(ii)   When oxygen was bubbled through a tank of sea water containing these organisms, the potassium ion concentration within the organisms increased.

Explain this effect.

_____

_____

_____   2

[Turn over

Marks

4. The diagram shows events occurring during the synthesis of a protein that is secreted from a cell.

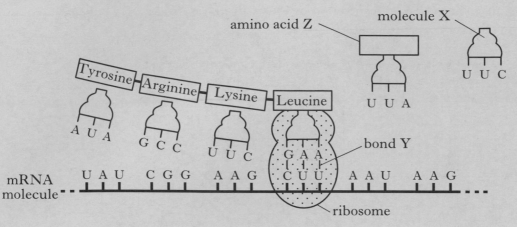

(*a*) (i) Name molecule X. _____ **1**

(ii) Name bond Y. _____ **1**

(*b*) What name is given to a group of three bases on mRNA that codes for an amino acid?

_____ **1**

(*c*) Give the sequence of DNA bases that codes for amino acid Z.

_____ **1**

(*d*) Describe the roles of the endoplasmic reticulum and the Golgi apparatus between the synthesis of the protein and its release from the cell.

Endoplasmic reticulum _____

_____ **1**

Golgi apparatus _____

_____ **1**

(*e*) The table contains some information about the structure and function of proteins.

Add information to the boxes to complete the table.

Protein	*Structure (Globular or Fibrous)*	*Function*
Cellulase		
Collagen		Structural protein in skin

2

Marks

5. (*a*) The graph shows the relationship between plant species diversity in grassland and grazing intensity by herbivores.

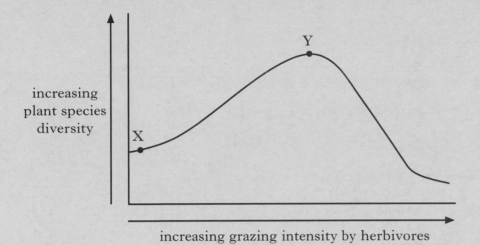

(i) Explain the effect on plant species diversity of increased grazing intensity by herbivores between X and Y on the graph.

_____

_____

_____ 2

(ii) What evidence is there that grassland contains plant species tolerant of grazing?

_____

_____ 1

(*b*) State **one** feature of some plant species that allows them to tolerate grazing by herbivores.

_____

_____ 1

[Turn over

6. Norway Spruce (*Picea abies*) is an evergreen species of tree with needle-like leaves, found in regions with extremely cold winters.

The rate of photosynthesis of the species is at its maximum during spring then decreases from June to December.

In an investigation, a sample of one-year-old seedlings was collected in each month from June to December.

For each sample of seedlings, the following measurements were made and averages calculated.

- Dry mass of whole seedlings
- Dry mass of roots only
- Starch content in needles
- Sugar content in needles

The results are shown in **Graphs 1** and **2**.

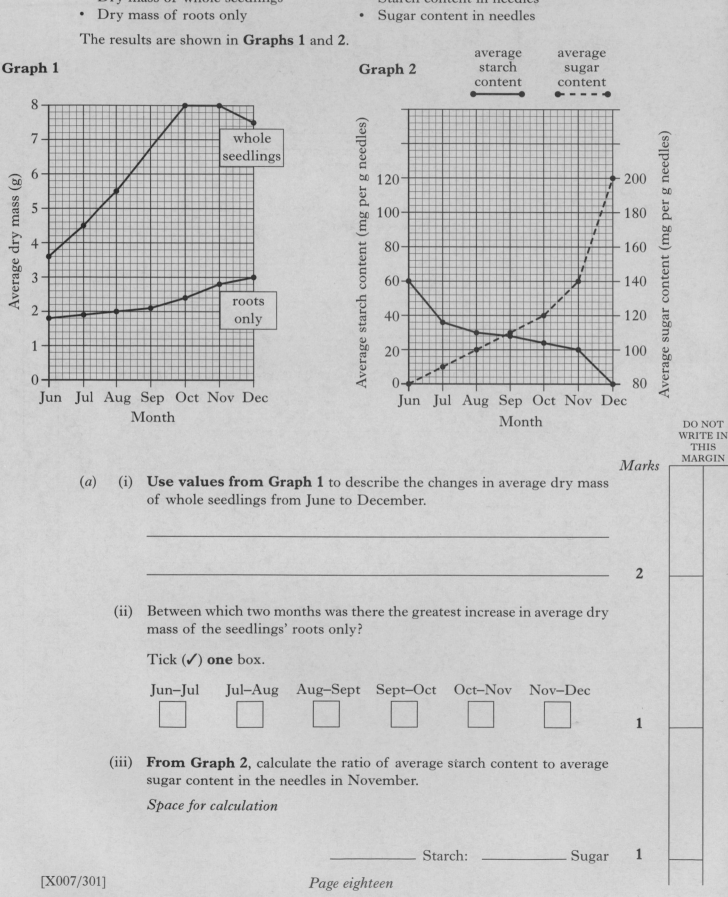

Graph 1

Graph 2

average starch content

average sugar content

Marks

(a) (i) **Use values from Graph 1** to describe the changes in average dry mass of whole seedlings from June to December.

_____

_____

2

(ii) Between which two months was there the greatest increase in average dry mass of the seedlings' roots only?

Tick (✓) **one** box.

Jun–Jul   Jul–Aug   Aug–Sept   Sept–Oct   Oct–Nov   Nov–Dec

☐   ☐   ☐   ☐   ☐   ☐

1

(iii) **From Graph 2**, calculate the ratio of average starch content to average sugar content in the needles in November.

Space for calculation

_____ Starch: _____ Sugar   1

Marks

6. (a) (continued)

(iv) **From Graph 2**, calculate the percentage decrease in average starch content in the needles between June and October.

Space for calculation

_____ % decrease   **1**

(b) Explain the decrease in average starch content in the needles between June and December.

_____

_____   **1**

(c) Raffinose is a sugar that prevents frost damage to needles.

The table shows the raffinose content of needles from the seedling samples.

Month	Raffinose content (mg per g of needles)
June	0
July	1
August	2
September	3
October	9
November	30
December	50

(i) What evidence is there that raffinose is not the only sugar present in the needles of Norway Spruce?

_____

_____   **1**

(ii) Suggest how the changing raffinose content of needles from June to December is of survival value to Norway Spruce.

_____

_____   **1**

[Turn over

Marks

7. The diagram shows stages in meiosis during which a mutation occurred and the effect of the mutation on the gametes produced.

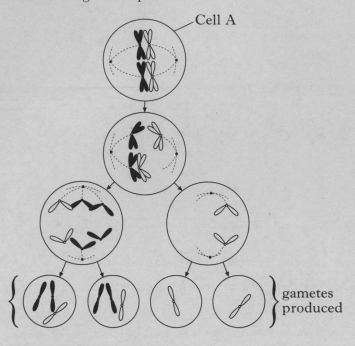

Cell A

gametes
produced

(*a*) (i) What name is given to cells such as cell A, that undergo meiosis?

_____ 1

(ii) Cell A contains two pairs of homologous chromosomes.

Apart from size and shape, state **one** similarity between homologous chromosomes.

_____ 1

(iii) This mutation has resulted in changes to the chromosome numbers in the gametes.

Name this type of mutation.

_____ 1

(iv) State whether the mutation has occurred in the first or second meiotic division and justify your choice.

Meiotic division _____

Justification _____

_____ 1

(v) State the expected haploid number of chromosomes in the gametes produced if this mutation **had not occurred**.

_____ 1

7. **(continued)**

Marks

(*b*) The diagram represents the sequence of bases on part of one strand of a DNA molecule.

Part of DNA molecule   T   G   A   A   C   T   G

The effects of two different gene mutations on the strand of DNA are shown below.

Gene mutation 1   T   T   G   A   A   C   T   G

Gene mutation 2   T   G   A   C   C   T   G

Complete the table by naming the type of gene mutation that has occurred in each case.

Gene Mutation	Name
1	
2	

2

[Turn over

Marks

8. The table shows information about three species of oak tree that have evolved from a common ancestor.

	Oak Species		
	Sessile Oak	Kermes Oak	Northern Red Oak
Leaf Shape	Rounded lobes	Sharp spines	Lobes with sharp spines
Growing Conditions	Mild and damp	Hot and dry	Cool and dry

(a) (i) The Oak species have evolved in ecological isolation.

State the importance of isolating mechanisms in the evolution of new species.

_____   1

(ii) Use the information to explain how the evolution of the Oak species illustrates adaptive radiation.

_____

_____   2

(b) The Kermes Oak grows to a maximum height of one metre.

Explain the benefit to this species of having leaves with sharp spines.

_____

_____   1

(c) To maintain genetic diversity, species must be conserved.

State **two** ways in which species can be conserved.

1 _____

2 _____   1

Marks

9. (*a*) The graph shows the body mass of a human male from birth until 22 years of age.

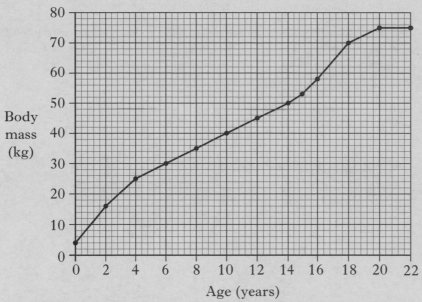

Age (years)

(i) Calculate the average yearly increase in body mass between age 6 and age 14.

Space for calculation

_____ kg   **1**

(ii) Explain how the activity of the pituitary gland could account for the growth pattern between 15 and 22 years of age shown on the graph.

_____

_____

_____   **2**

(*b*) The diagram shows the role of the pituitary gland in the secretion of a hormone from the thyroid gland.

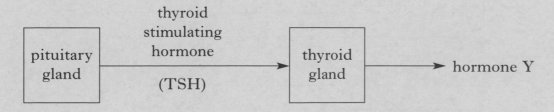

Name hormone Y and describe its role in the control of growth and development.

Hormone Y_____   **1**

Role _____   **1**

Marks

10. An experiment was carried out to investigate the effect of gibberellic acid (GA) on the growth of dwarf pea plants. GA can be absorbed by leaves.

Six identical pea plants were placed in pots containing 100 g of soil. The leaves of each were sprayed with an equal volume of water containing a different mass of GA. The soil in each pot received 20 cm³ water each day and the plants were continuously exposed to equal light intensity from above.

After seven days the stem height of each plant was measured and the percentage increase in stem height calculated.

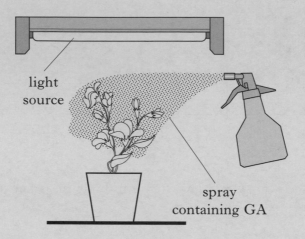

(*a*) (i) Identify **two** variables, not already mentioned, that should have been controlled to ensure the experimental procedure was valid.

Variable 1 _____   1

Variable 2 _____   1

(ii) State **one** way in which the experimental procedure could be improved to increase the reliability of the results.

_____

_____   1

(*b*) The results of the experiment are shown in the table.

Mass of GA applied (micrograms)	Percentage increase in stem height (%)
0·01	90
0·03	120
0·05	160
0·08	240
0·10	320
0·11	350

DO NOT
WRITE IN
THIS
MARGIN

Marks

10. **(b)** **(continued)**

(i) On the grid provided, complete the line graph to show the percentage increase in stem height against the mass of GA applied.

Use an appropriate scale to fill most of the grid.

(Additional graph paper, if required, will be found on page 36.)

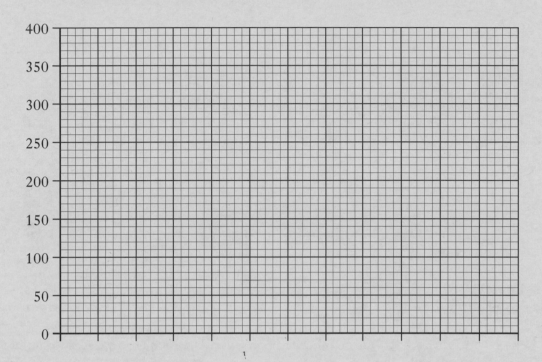

2

(ii) Another pea plant was treated in the same way, using a water spray containing 0·12 micrograms of GA. Predict the percentage increase in stem height of this plant after seven days.

_____ percentage increase   1

(c) Explain why the method of application of GA could lead to errors in the results.

_____

_____   1

[Turn over

Marks

11. (*a*) The bacterium *Escherichia coli* can control its lactose metabolism.

Complete **all** boxes in the table to show whether each statement is true (T) or false (F) if lactose is present or absent in the medium in which *E. coli* is growing.

Statement	*Lactose present*	*Lactose absent*
Regulator gene produces the repressor molecule		
Repressor molecule binds to inducer		
Repressor molecule binds to operator		
Structural gene switched on	T	F

2

(*b*) Part of a metabolic pathway involving the amino acid phenylalanine is shown in the diagram.

digestion of protein in diet ——→ **phenylalanine** ——enzyme A——→ **tyrosine** ——enzyme B——→ **other compounds**

Phenylketonuria (PKU) is an inherited condition in which enzyme A is either absent or does not function.

(i) Predict the effect on the concentrations of phenylalanine and tyrosine if enzyme A is absent.

Phenylalanine _____ 1

Tyrosine _____ 1

(ii) PKU is caused by a mutation of the gene that codes for enzyme A.

Explain how a mutation of a gene can cause the production of an altered enzyme.

_____

_____

_____ 2

Marks

12. An experiment was set up to investigate the effect of photoperiod on flowering in *Chrysanthemum* plants. Four plants A, B, C and D were exposed to different periods of light and dark in 24 hours. This was repeated every day for several weeks and the effects on flowering noted.

The periods of light and dark and their effects on flowering are shown in the diagram.

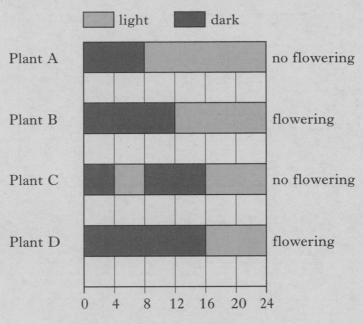

(a) From the information given, identify the conditions required for flowering in *Chrysanthemum* plants. Justify your answer.

Conditions _____

Justification _____

_____ 1

(b) Flowering in response to photoperiod ensures plants within a population flower at the same time. Explain how this enables genetic variation to be maintained.

_____

_____ 1

(c) Mammals also show photoperiodism.

Describe how one type of mammal behaviour can be affected by photoperiod.

_____

_____ 1

[Turn over

Marks

13. The homeostatic control of blood glucose concentration carried out by the human liver is shown on the diagram.

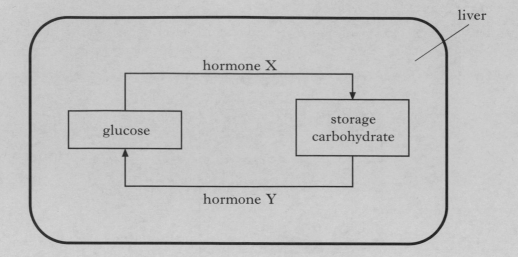

(a) Name the storage carbohydrate found in the liver.

_____

1

(b) (i) Name hormones X and Y.

Hormone X _____

Hormone Y _____

1

(ii) Name the organ that produces hormones X and Y.

_____

1

(iii) Explain how negative feedback is involved in the homeostatic control of blood glucose concentration.

_____

_____

_____

2

Marks

14. (*a*) The diagram shows some plant communities present at various time intervals on farmland cleared of vegetation by a fire.

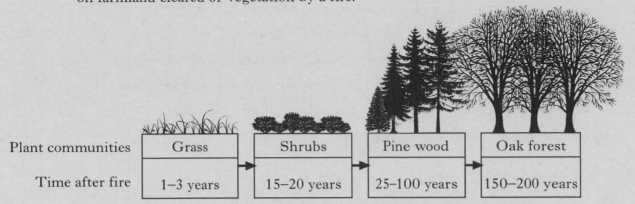

Plant communities	Grass	Shrubs	Pine wood	Oak forest
Time after fire	1–3 years	15–20 years	25–100 years	150–200 years

(i) State the term used to describe this sequence of plant communities.

_____ 1

(ii) Give a reason to explain why the shrub community is able to replace the grass community after 15 years.

_____ 1

(iii) Oak forest is the climax community in this sequence.
Describe a feature of a climax community.

_____ 1

(*b*) The grid shows factors that can influence population change.

A competition	B predation	C rainfall
D disease	E temperature	F food supply

(i) Use **all** the letters from the grid to complete the table to show which factors are density dependent and which are density independent.

Density dependent	Density independent

2

(ii) **Underline** one alternative in each pair to make the sentences correct.

As population density $\left\{\begin{array}{l}\text{increases,}\\\text{decreases,}\end{array}\right\}$ the effect of density dependent factors increases.

As a result, the population density then $\left\{\begin{array}{l}\text{increases}\\\text{decreases}\end{array}\right\}$. 1

[Turn over for Section C on *Page thirty*

SECTION C

Both questions in this section should be attempted.

Note that each question contains a choice.

Questions 1 and 2 should be attempted on the blank pages which follow.

Supplementary sheets, if required, may be obtained from the invigilator.

All answers must be written clearly and legibly in ink.

Labelled diagrams may be used where appropriate.

Marks

1. Answer **either** A **or** B.

    A. Give an account of respiration under the following headings:

    (i)   glycolysis;   **5**

    (ii) the Krebs (Citric acid) cycle.   **5**

    OR   **(10)**

    B. Give an account of cellular defence mechanisms in animals under the following headings:

    (i)   phagocytosis;   **4**

    (ii) antibody production and tissue rejection.   **6**

                                                                         (10)

In question 2, ONE mark is available for coherence and ONE mark is available for relevance.

2. Answer **either** A **or** B.

    A. Give an account of the problems of osmoregulation in freshwater bony fish and outline their adaptations to overcome these problems.   **(10)**

    OR

    B. Give an account of obtaining food in animals by reference to co-operative hunting, dominance hierarchy, and territorial behaviour.   **(10)**

[END OF QUESTION PAPER]

DO NOT
WRITE IN
THIS
MARGIN

SPACE FOR ANSWERS

[Turn over

SPACE FOR ANSWERS

DO NOT WRITE IN THIS MARGIN

SPACE FOR ANSWERS

[Turn over

SPACE FOR ANSWERS

Page thirty-four

DO NOT WRITE IN THIS MARGIN

SPACE FOR ANSWERS

DO NOT
WRITE IN
THIS
MARGIN

SPACE FOR ANSWERS

ADDITIONAL GRAPH PAPER FOR QUESTION 10(*b*)

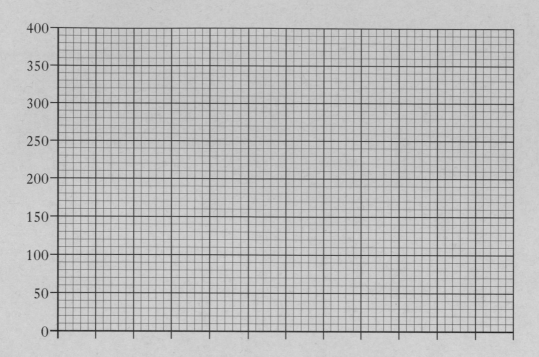

[BLANK PAGE]

FOR OFFICIAL USE

Total for
Sections
B and C

X007/301

NATIONAL
QUALIFICATIONS
2008

TUESDAY, 27 MAY
1.00 PM – 3.30 PM

BIOLOGY
HIGHER

Fill in these boxes and read what is printed below.

Full name of centre

Town

Forename(s)

Surname

Date of birth
Day Month Year   Scottish candidate number   Number of seat

SECTION A—Questions 1–30 (30 marks)

Instructions for completion of Section A are given on page two.

For this section of the examination you must use an **HB pencil**.

SECTIONS B AND C (100 marks)

1   (a)   All questions should be attempted.

    (b)   It should be noted that in **Section C** questions 1 and 2 each contain a choice.

2   The questions may be answered in any order but all answers are to be written in the spaces provided in this answer book, **and must be written clearly and legibly in ink**.

3   Additional space for answers will be found at the end of the book. If further space is required, supplementary sheets may be obtained from the invigilator and should be inserted inside the **front** cover of this book.

4   The numbers of questions must be clearly inserted with any answers written in the additional space.

5   Rough work, if any should be necessary, should be written in this book and then scored through when the fair copy has been written. If further space is required a supplementary sheet for rough work may be obtained from the invigilator.

6   Before leaving the examination room you must give this book to the invigilator. If you do not, you may lose all the marks for this paper.

Read carefully

1   Check that the answer sheet provided is for **Biology Higher (Section A)**.

2   For this section of the examination you must use an **HB pencil**, and where necessary, an eraser.

3   Check that the answer sheet you have been given has **your name**, **date of birth**, **SCN** (Scottish Candidate Number) and **Centre Name** printed on it.

    Do not change any of these details.

4   If any of this information is wrong, tell the Invigilator immediately.

5   If this information is correct, **print** your name and seat number in the boxes provided.

6   The answer to each question is **either** A, B, C or D. Decide what your answer is, then, using your pencil, put a horizontal line in the space provided (see sample question below).

7   There is **only one correct** answer to each question.

8   Any rough working should be done on the question paper or the rough working sheet, **not** on your answer sheet.

9   At the end of the exam, put the **answer sheet for Section A inside the front cover of this answer book**.

Sample Question

The apparatus used to determine the energy stored in a foodstuff is a

A   calorimeter

B   respirometer

C   klinostat

D   gas burette.

The correct answer is **A**—calorimeter. The answer **A** has been clearly marked in **pencil** with a horizontal line (see below).

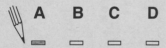

Changing an answer

If you decide to change your answer, carefully erase your first answer and using your pencil fill in the answer you want. The answer below has been changed to **D**.

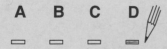

SECTION A

All questions in this section should be attempted.

Answers should be given on the separate answer sheet provided.

1. The following statements relate to respiration and the mitochondrion.

   1 Glycolysis takes place in the mitochondrion.

   2 The mitochondrion has two membranes.

   3 The rate of respiration is affected by temperature.

   Which of the above statements are correct?

   A 1 and 2

   B 1 and 3

   C 2 and 3

   D All of them

2. The anaerobic breakdown of glucose splits from the aerobic pathway of respiration

   A after the formation of pyruvic acid

   B after the formation of acetyl-CoA

   C after the formation of citric acid

   D at the start of glycolysis.

3. Phagocytes contain many lysosomes so that

   A enzymes which destroy bacteria can be stored

   B toxins from bacteria can be neutralised

   C antibodies can be released in response to antigens

   D bacteria can be engulfed into the cytoplasm.

4. After an animal cell is immersed in a hypotonic solution it will

   A burst

   B become turgid

   C shrink

   D become flaccid.

5. Which of the following proteins has a fibrous structure?

   A Pepsin

   B Amylase

   C Insulin

   D Collagen

6. The following cell components are involved in the synthesis and secretion of an enzyme.

   1 Golgi apparatus

   2 Ribosome

   3 Cytoplasm

   4 Endoplasmic reticulum

   Which of the following identifies correctly the route taken by an amino acid molecule as it passes through these cell components?

   A 3 2 1 4

   B 2 4 3 1

   C 3 2 4 1

   D 3 4 2 1

[Turn over

7. The graphs show the effect of various factors on the rate of uptake of chloride ions by discs of carrot tissue from their surrounding solution.

1

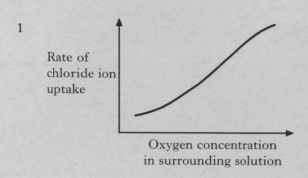

Rate of chloride ion uptake

Oxygen concentration in surrounding solution

2

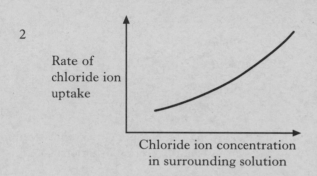

Rate of chloride ion uptake

Chloride ion concentration in surrounding solution

3

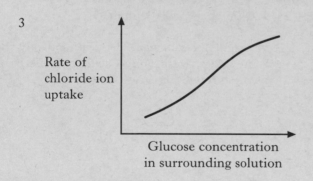

Rate of chloride ion uptake

Glucose concentration in surrounding solution

Which graphs support the hypothesis that chloride ion uptake by carrot tissue involves active transport?

A 1 and 2 only

B 1 and 3 only

C 2 and 3 only

D 1, 2 and 3

8. The R_f value of a pigment can be calculated as follows:

$$R_f = \frac{\text{distance travelled by pigment from origin}}{\text{distance travelled by solvent from origin}}$$

The diagram shows a chromatogram in which four chlorophyll pigments have been separated.

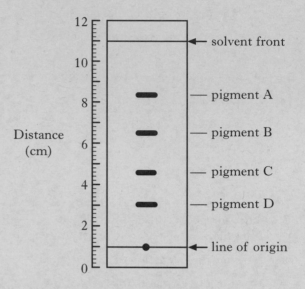

Which pigment has an R_f value of 0·2?

9. The following steps occur during the replication of a virus.

1 alteration of host cell metabolism

2 production of viral protein coats

3 replication of viral nucleic acid

In which sequence do these events occur?

A 1, 3, 2

B 1, 2, 3

C 2, 1, 3

D 3, 1, 2

10. The graphs below show the effect of two injections of an antigen on the formation of an antibody.

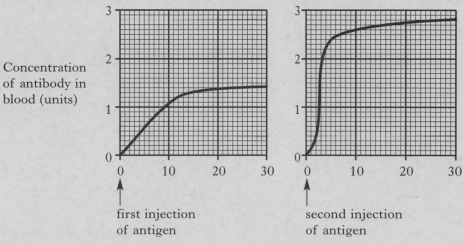

first injection
of antigen

second injection
of antigen

Time in days

The concentration of antibodies is measured 25 days after each injection. The effect of the second injection is to increase the concentration by

A   1%

B   25%

C   50%

D   100%.

11. The diagram shows a stage of meiosis.

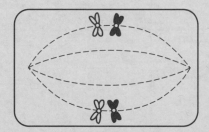

Which of the following diagrams shows the next stage in meiosis?

A
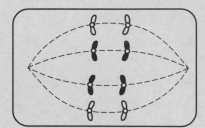

B
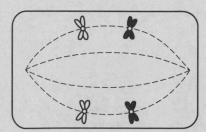

C
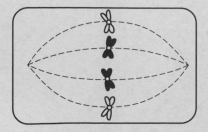

D
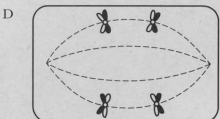

12. Cystic fibrosis is a genetic condition caused by an allele that is not sex-linked. A child is born with cystic fibrosis despite neither parent having the condition. The parents are going to have a second child.

    What is the chance that the second child will have cystic fibrosis?

    A 75%

    B 67%

    C 50%

    D 25%

13. A sex-linked condition in humans is caused by a recessive allele.

    An unaffected man and a carrier woman produce a son.

    What is the chance that he will be unaffected?

    A 1 in 1

    B 1 in 2

    C 1 in 3

    D 1 in 4

14. A new species of organism is considered to have evolved when its population

    A is isolated from the rest of the population by a geographical barrier

    B shows increased variation due to mutations

    C can no longer interbreed successfully with the rest of the population

    D is subjected to increased selection pressure in its habitat.

15. The melanic variety of the peppered moth became common in industrial areas of Britain following the increase in the production of soot during the industrial revolution.

    The increase in the melanic variety was due to

    A melanic moths migrating to areas which gave the best camouflage

    B a change in selection pressure

    C an increase in the mutation rate

    D a change in the prey species taken by birds.

16. Which of the following is true of freshwater fish?

    A The kidneys contain few small glomeruli.

    B The blood filtration rate is high.

    C Concentrated urine is produced.

    D The chloride secretory cells actively excrete excess salts.

17. Which of the following is a behavioural adaptation used by some mammals to survive in hot deserts?

    A Dry mouth and nasal passages

    B High levels of anti-diuretic hormone in the blood

    C Very long kidney tubules

    D Nocturnal habit

18. Which line in the table below correctly identifies the effect of the state of the guard cells on the opening and closing of stomata?

	State of guard cells	Stomata open/closed
A	flaccid	open
B	plasmolysed	open
C	flaccid	closed
D	turgid	closed

19. In an animal, habituation has taken place when a

    A harmful stimulus ceases to produce a response

    B harmful stimulus always produces an identical response

    C harmless stimulus ceases to produce a response

    D harmless stimulus always produces an identical response.

20. The table below shows the rate of production of urine by a salmon in both fresh and sea water.

	Rate of urine production (cm^3/kg of body mass/hour)
In fresh water	5·0
In sea water	0·5

After transfer from the sea to fresh water, the volume of urine produced by a 2·5 kg salmon over a one hour period would have increased by

A   4·50 cm^3

B   5·50 cm^3

C   11·25 cm^3

D   12·50 cm^3.

21. A 30 g serving of a breakfast cereal contains 1·5 mg of iron. Only 25% of this iron is absorbed into the bloodstream.

If a pregnant woman requires a daily uptake of 6 mg of iron, how much cereal would she have to eat each day to meet this requirement?

A   60 g

B   120 g

C   240 g

D   480 g

22. Which of the following graphs represents the growth pattern of a locust?

A
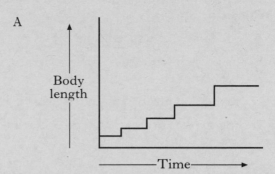

B
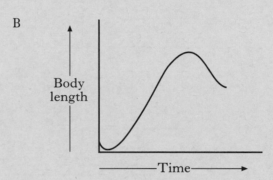

C
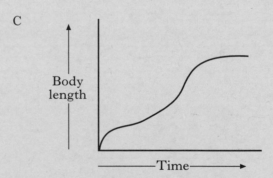

D
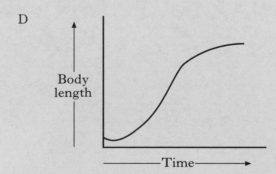

[Turn over

23. Gene expression in cells results in the synthesis of specific proteins. The process of transcription involved in the synthesis of a protein is the

   A   production of a specific mRNA

   B   processing of a specific mRNA on the ribosomes

   C   replication of DNA in the nucleus

   D   transfer of amino acids to the ribosomes.

24. Hormones P and Q are involved in the control of growth and metabolism.

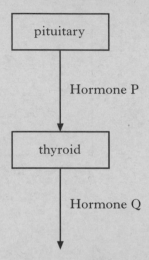

Control of growth and metabolism

Which line in the table below correctly identifies hormones P and Q?

	Hormone P	Hormone Q
A	TSH	growth hormone
B	thyroxine	TSH
C	TSH	thyroxine
D	growth hormone	TSH

25. Temperature control mechanisms in the skin of mammals are stimulated by

   A   nerve impulses from the pituitary gland

   B   nerve impulses from the hypothalamus

   C   hormonal messages from the hypothalamus

   D   hormonal messages from the pituitary gland.

26. Which of the following is **not** an effect of IAA?

   A   Increased stem elongation

   B   Fruit formation

   C   Inhibition of leaf abscission

   D   Initiation of germination

27. List P gives reasons why population monitoring may be carried out.

List Q gives three species whose populations are monitored by scientists.

List P	List Q
1 Valuable food resource	W Stonefly
2 Endangered species	X Humpback Whale
3 Indicator species	Y Haddock

Which line in the table below correctly matches reasons from **List P** with species from **List Q**?

	Reasons		
	1	2	3
A	W	X	Y
B	Y	W	X
C	X	Y	W
D	Y	X	W

28. The graph below records the body temperature of a woman during an investigation in which her arm was immersed in water.

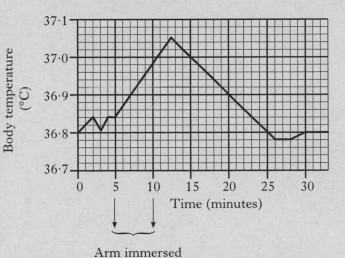

By how much did the temperature of her body vary during the 30 minutes of the investigation?

A   0·25 °C

B   0·27 °C

C   2·5 °C

D   2·7 °C

29. The graph below shows the variation in numbers of a predator and its prey recorded over a ten week period.

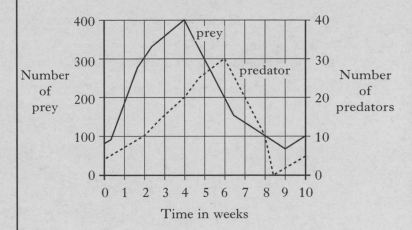

In which week is the prey to predator ratio the largest?

A   week 2

B   week 4

C   week 6

D   week 8

30. The graph below shows the length of a human fetus before birth.

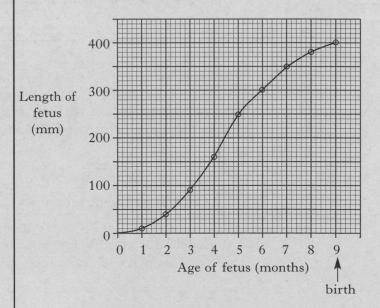

What is the percentage increase in length of the fetus during the 4 months before birth?

A   33·3%

B   37·5%

C   60·0%

D   150%

Candidates are reminded that the answer sheet MUST be returned INSIDE the front cover of this answer book.

Marks

SECTION B

All questions in this section should be attempted.

All answers must be written clearly and legibly in ink.

1. The diagram shows a chloroplast from a palisade mesophyll cell.

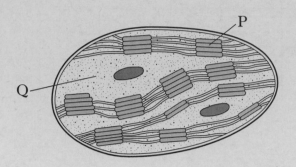

(a) Name regions P and Q.

P _____

Q _____   **1**

(b)   (i)   Mark an X on the diagram to show the location of chlorophyll molecules.   **1**

    (ii)   As well as chlorophyll, plants have other photosynthetic pigments.

       State the benefit to plants of having these other pigments.

       _____

       _____   **1**

(c)   (i)   Name **one** product of the light dependent stage of photosynthesis which is required for the carbon fixation stage (Calvin cycle).

       _____   **1**

    (ii)   The table shows some substances involved in the carbon fixation stage of photosynthesis.

       Complete the table by inserting the number of carbon atoms present in one molecule of each substance.

Substance	*Number of carbon atoms in one molecule*
Glucose	
Carbon dioxide	
Glycerate phosphate (GP)	
Ribulose bisphosphate (RuBP)	

  2

Marks

1. **(continued)**

   (*d*) The graph below shows the effect of increasing light intensity on the rate of photosynthesis at different carbon dioxide concentrations and temperatures.

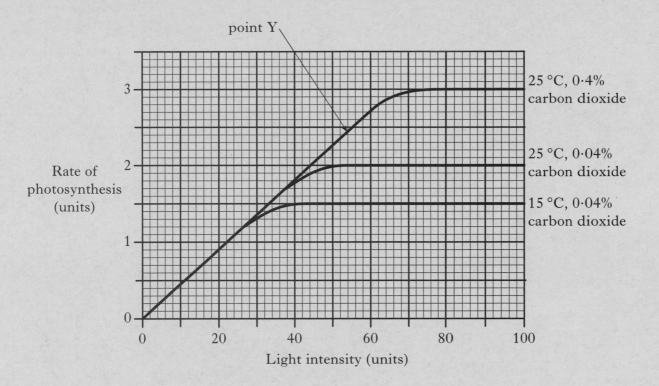

   (i) Identify the factor limiting the rate of photosynthesis at point Y on the graph.

   _____ 1

   (ii) From the graph, identify the factor that has the greatest effect in increasing the rate of photosynthesis at a light intensity of 80 units.

   Justify your answer.

   Factor _____

   Justification _____

   _____

   _____ 1

   [Turn over

Page eleven

Marks

2. The diagram shows a human liver cell and a magnified section of its plasma membrane.

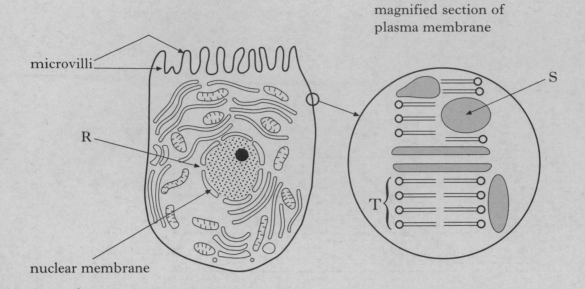

magnified section of
plasma membrane

microvilli

R

nuclear membrane

S

T

(a) (i) Identify molecules S and T.

S _____   1

T _____   1

(ii) A pore in the nuclear membrane is shown by label R.

Describe the importance of these pores in protein synthesis.

_____

_____   1

(iii) What evidence in the diagram suggests that this cell produces large quantities of ATP?

_____   1

Marks

2.   **(continued)**

(*b*)   Some liver cells take up glucose from the blood by the process of diffusion.

(i)   Describe this process.

_____

_____   1

(ii)   Suggest a reason for the presence of microvilli in liver cells as shown in the diagram.

_____

_____

_____   2

(iii)   Glucose taken up by liver cells can be converted into a storage carbohydrate.

Name this carbohydrate.

_____   1

[Turn over

Marks

3. Fat can be used as an alternative respiratory substrate. The diagram shows the breakdown of fat in an athlete's muscle cells during the final stages of a marathon race.

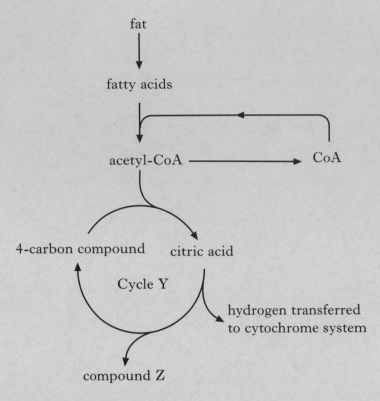

(*a*) Name a respiratory substrate, other than fat, which can be used by muscle cells.

_____   1

(*b*) (i) Name cycle Y and compound Z.

Cycle Y _____

Compound Z _____   1

(ii) Name the carrier that accepts and transfers hydrogen to the cytochrome system.

_____   1

(*c*) Describe the role of oxygen in aerobic respiration.

_____

_____   1

Marks

3. **(continued)**

(d) During an 800 metre race, an athlete's muscle cells may respire anaerobically to produce ATP.

    (i) State **one** other metabolic product of anaerobic respiration in muscle cells.

       _____   1

    (ii) Where in a cell does anaerobic respiration occur?

       _____   1

    (iii) Describe the importance of ATP to cells.

       _____

       _____   1

[Turn over

Marks

4. (*a*) Decide if each of the following statements about DNA replication is **True** or **False** and tick (✓) the appropriate box.

If the statement is False, write the correct word in the correction box to replace the word underlined in the statement.

Statement	True	False	Correction
During DNA replication hydrogen bonds between bases break.			
During the formation of a new DNA molecule, base pairing is followed by bonding between deoxyribose and bases.			
As a result of DNA replication, the DNA content of a cell is halved.			

2

(*b*) Free DNA nucleotides are needed for DNA replication.

Name **one** other substance that is needed for DNA replication.

_____

1

(*c*) A single strand of a DNA molecule has 6000 nucleotides of which 24% are adenine and 18% are cytosine.

(i) Calculate the combined percentage of thymine and guanine bases on the same DNA strand.

Space for calculation

_____ % 1

(ii) How many guanine bases would be present on the complementary strand of this DNA molecule?

Space for calculation

_____ bases 1

Marks

5. Ponderosa pine trees produce resin following damage to their bark.

In an investigation, three individual pine trees were chosen from areas with different population densities. Each tree was damaged by having a hole bored through its bark.

Measurements of resin production from each hole following this damage are shown in the table.

Population density (Number of trees per hectare)	Volume of resin produced in the first day (cm³)	Duration of resin flow (days)	Total volume of resin produced (cm³)
2	8·3	7·0	29·3
10	0·8	4·8	2·9
50	0·6	4·6	2·8

(a) (i) Describe how population density affects the total volume of resin produced.

_____

_____

_____   2

(ii) Calculate the average resin flow per day at a population density of 2 trees per hectare **after the first day**.

Space for calculation

_____ cm³ per day   1

(b) Explain how resin production protects trees.

_____

_____   1

[Turn over

Marks

6. An investigation was carried out to compare the rates of water loss from tree species during winter when soil water availability is low.

The table shows information about the tree species involved.

Tree species	Leaf type	Leaves lost in winter
cherry laurel	broad	no
white oak	broad	yes
loblolly pine	needle-like	no

One year old trees of each species were grown outside in identical environmental conditions during winter. The average rate of water loss from each species was measured every tenth day over a 70 day period.

The results are shown in **Graph 1**.

Graph 1

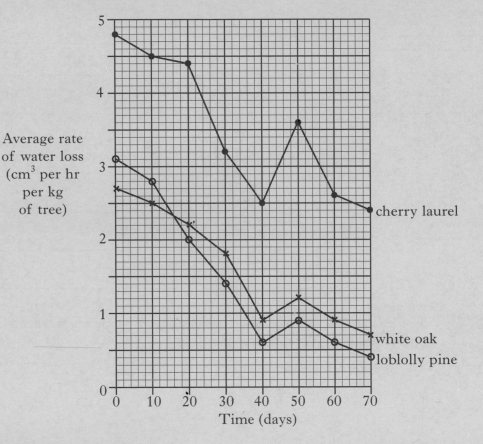

(a) (i) **Use values from Graph 1** to describe the changes in rate of water loss from loblolly pine over the 70 day period.

_____

_____

_____   2

(ii) Calculate the percentage decrease in rate of water loss from cherry laurel between day 0 and day 50.

Space for calculation

_____ %   1

6. **(a)** **(continued)**

(iii) **From Graph 1** express, as the simplest whole number ratio, the rates of water loss from white oak and cherry laurel on day 20.

_____white oak : _____cherry laurel

1

(iv) Using the information from the table **and** from Graph 1, suggest the advantage to the white oak of losing its leaves in winter.

Justify your answer.

Advantage _____

Justification _____

_____

_____

2

(b) In a further investigation, the effect of air temperature on the average rate of water loss from loblolly pine was measured.

The results are shown in **Graph 2**.

Graph 2

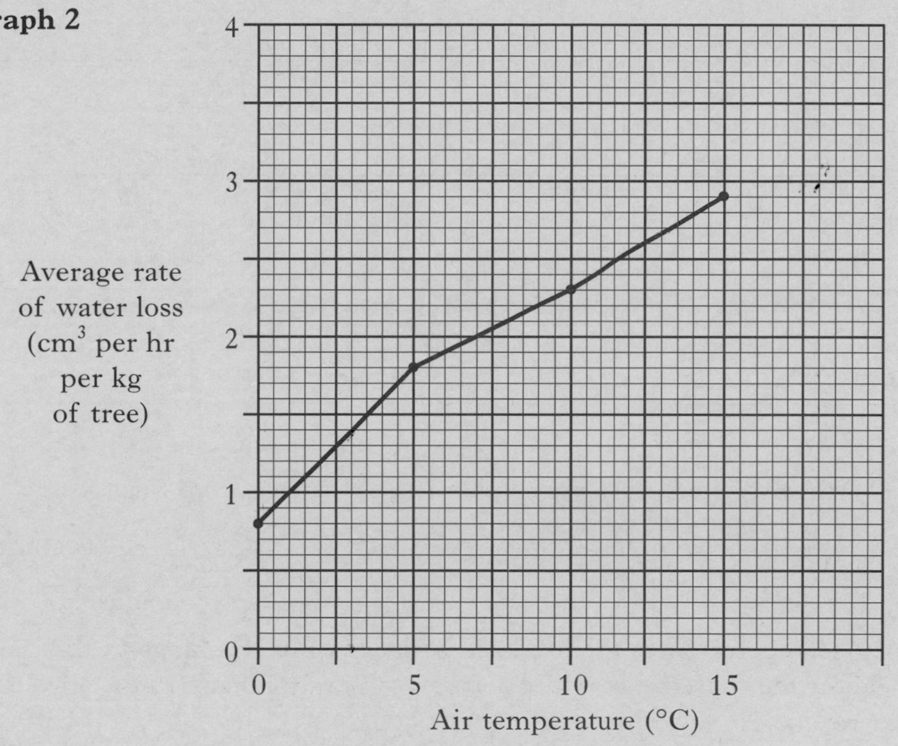

Average rate of water loss (cm³ per hr per kg of tree)

Air temperature (°C)

(i) Use the information from **Graphs 1 and 2** to suggest the air temperature on day 30 of the investigation.

_____ °C   1

(ii) Predict the rate of water loss from loblolly pine at an air temperature of 18 °C.

_____cm³ per hour per kg of tree.   1

(iii) Apart from air temperature and soil water availability, state **one** factor which can affect water loss from trees.

_____   1

Marks

7. In horses, coat colour is determined by two genes. The allele for black coat (**B**) is dominant to the allele for chestnut coat (**b**). The allele for grey coat (**G**) is dominant to the allele for non-grey coat (**g**).

Horses with the allele **G** are **always** grey.

A male with the genotype **GgBb** was crossed with a female with the genotype **ggBb**.

(*a*) (i) State the phenotype of each parent.

Male _____

Female _____   1

(ii) Complete the grid by adding the genotypes of:

1 the male and female gametes;   1

2 the possible offspring.   1

	Male gametes			
Female gametes				

(iii) Give the expected phenotype ratio of the offspring from this cross.

_____ Grey : _____ Black : _____ Chestnut   1

(*b*) A further gene determines the presence of large white markings in the coat. The allele for the presence of white markings (**T**) is dominant to the allele for their absence (**t**).

A breeder found that a male horse with white markings always produced offspring with white markings when crossed with a female of any phenotype.

Explain this observation in terms of the genotype of this male horse.

_____

_____   1

DO NOT
WRITE IN
THIS
MARGIN

Marks

8. Grey wolves hunt in packs. Their prey includes a variety of large herbivores.

(*a*) (i) Name the hunting method used by wolves and state **one** advantage of this method.

Name _____ 1

Advantage _____

_____ 1

(ii) Following the capture of prey, higher ranking wolves feed first.

State the term which describes this type of social organisation.

_____ 1

(iii) Wolf packs occupy territories ranging from 80 to 1500 km^2.

1 Describe **one** advantage to the wolf pack of occupying a territory.

_____

_____ 1

2 Suggest **one** factor that could influence the size of a territory occupied by a wolf pack.

_____ 1

(*b*) The grey wolf was once common in North America but is now an endangered species in many areas.

Following steps to conserve the species, wolf numbers in one wildlife reserve increased from 31 to 683 individuals during an eight year period.

(i) Calculate the average yearly increase in wolf numbers during this period.

Space for calculation

_____ per year 1

(ii) Other than wildlife reserves, describe **one** method used to conserve endangered species.

_____ 1

[Turn over

9. Beech trees have two types of leaf. Sun leaves are exposed to high light intensities for most of the day and shade leaves are usually overshadowed by sun leaves.

The rates of carbon dioxide exchange at different light intensities were measured for sun leaves and shade leaves from one beech tree.

The results are shown on the graph.

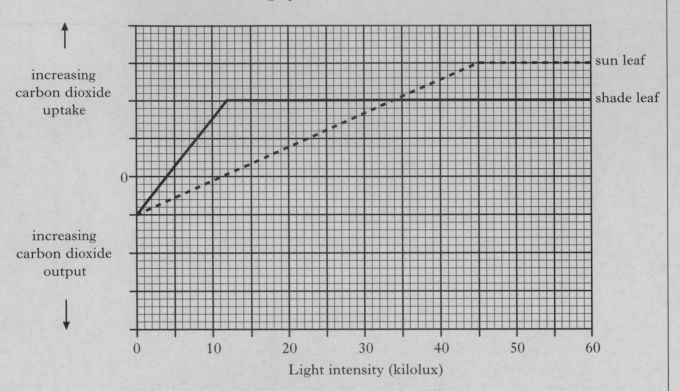

(a) State the light intensity at which the shade leaves reach their compensation point.

_____ kilolux   **1**

(b) Explain why having shade leaves is an advantage to a beech tree.

_____

_____   **1**

DO NOT
WRITE IN
THIS
MARGIN

Marks

10. Camels live in deserts where temperatures rise to 50 °C during the day and fall to minus 10 °C at night. The graph shows how the body temperature of a camel varied over a three day period.

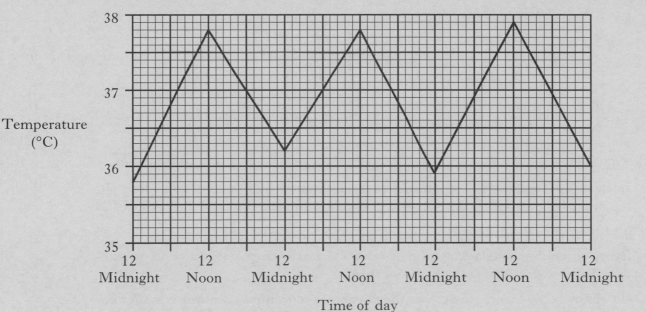

(*a*) From the information given, what evidence is there that camels obtain heat from their own metabolism?

_____

_____ 1

(*b*) What term is used for animals that obtain most of their body heat from their own metabolism?

_____ 1

[Turn over

11. In an investigation into the effect of potassium on barley root growth, twelve *Marks*
containers were set up as shown.

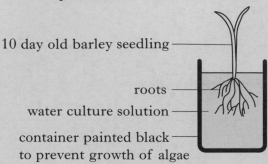

10 day old barley seedling

roots

water culture solution

container painted black
to prevent growth of algae

The water culture solution provided all the elements needed for normal growth.

In six of the containers, the potassium concentration was 2 micromoles per litre. In the other six containers, the potassium concentration was 5 millimoles per litre.

The containers were kept at 20 °C and in constant light intensity.

Every three days, the roots from one container at each potassium concentration were harvested and their dry mass measured.

(a) How many times greater was the potassium concentration in the 5 millimoles per litre solution than in the 2 micromoles per litre solution?

1 millimole per litre = 1000 micromoles per litre

Space for calculation

_____ times   **1**

(b) (i) Identify **one** variable, not already described, that should be kept constant.

_____   **1**

(ii) Suggest **one** advantage of growing the seedlings in water culture solutions rather than soil.

_____   **1**

(iii) Complete the table to give the reasons for each experimental procedure.

Experimental procedure	Reason
paint containers black to prevent growth of algae	
measure dry mass rather than fresh mass of roots	

2

(c) The results of the investigation are shown in the table.

Time (days)	Dry mass of roots (mg)	
	Potassium concentration 2 micromoles per litre	Potassium concentration 5 millimoles per litre
3	1	1
6	5	6
9	8	10
12	11	14
15	16	22
18	22	44

The results for the seedlings grown in 5 millimoles potassium per litre solution are shown on the graph.

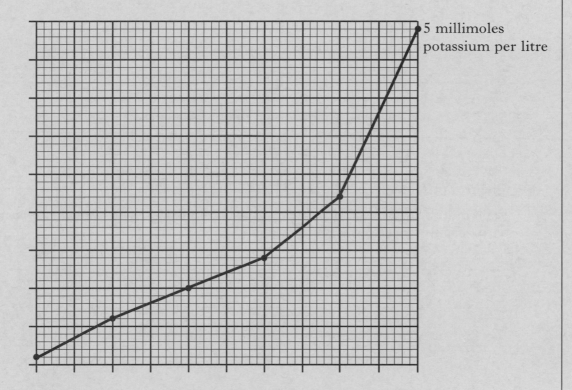

5 millimoles potassium per litre

Complete the graph by:

(i) adding the scale and label to each axis;

1

(ii) presenting the results for the 2 micromoles potassium per litre solution **and** labelling the line.

1

(Additional graph paper, if required, will be found on page 36.)

(d) In a further experiment, bubbling oxygen through the water culture solutions was observed to increase the uptake of potassium by the barley roots.

Explain this observation.

_____

_____

_____

2

Marks

DO NOT
WRITE IN
THIS
MARGIN

Marks

12. The diagram shows a section through a three year old hawthorn twig with annual rings.

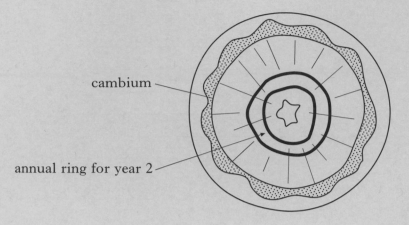

cambium

annual ring for year 2

(*a*)   (i)   State the function of cambium.

_____ **1**

    (ii)   Name the tissue of which annual rings are composed.

_____ **1**

(*b*)   The tree suffered an infestation of leaf-eating caterpillars during year 2.

Explain how an infestation of leaf-eating caterpillars could account for the narrow appearance of this annual ring.

_____

_____

_____ **2**

Marks

13. The flow chart shows part of the homeostatic control of water concentration in human blood.

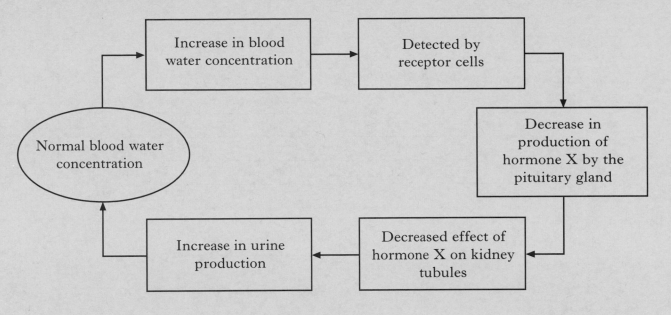

(*a*) (i) Suggest a reason for the increase in blood water concentration.

_____   1

(ii) State the location of the receptor cells.

_____   1

(iii) Name hormone X and state its effect on the kidney tubules.

Name _____   1

Effect _____

_____   1

(*b*) Control of water concentration in human blood involves negative feedback.
Explain what is meant by negative feedback.

_____

_____

_____   2

[Turn over

Marks

14. Frequency of mating in a population of wild goats was observed from June to November.

The results are shown in the table.

Month	Average number of hours of light per day	Frequency of mating 0 = no mating + = occasional mating ++ = frequent mating
June	19	0
July	17	0
August	15	0
September	13	+
October	11	++
November	9	++

(a) Using the information given, identify the trigger stimulus which results in mating of goats.

_____   1

(b) Young goats are born 5 months after mating.

Explain how the pattern of mating frequency shown increases the survival rate of the offspring.

_____

_____

_____   2

(c) What general term is used to describe the effect of light on the timing of breeding in mammals such as goats?

_____   1

Marks

SECTION C

Both questions in this section should be attempted.

Note that each question contains a choice.

Questions 1 and 2 should be attempted on the blank pages which follow.

Supplementary sheets, if required, may be obtained from the invigilator.

All answers must be written clearly and legibly in ink.

Labelled diagrams may be used where appropriate.

1. Answer **either** A **or** B.

   A. Write notes on:

   (i) the control of lactose metabolism in *E. coli*;   6

   (ii) phenylketonuria in humans.   4

   (10)

   OR

   B. Write notes on population change under the following headings:

   (i) the influence of density dependent factors;   5

   (ii) succession in plant communities.   5

   (10)

In question 2, ONE mark is available for coherence and ONE mark is available for relevance.

2. Answer **either** A **or** B.

   A. Give an account of gene mutations and mutagenic agents.   **(10)**

   OR

   B. Give an account of somatic fusion in plants and genetic engineering in bacteria.   **(10)**

[END OF QUESTION PAPER]

SPACE FOR ANSWERS

SPACE FOR ANSWERS

[Turn over

DO NOT WRITE IN THIS MARGIN

SPACE FOR ANSWERS

SPACE FOR ANSWERS

DO NOT WRITE IN THIS MARGIN

SPACE FOR ANSWERS

SPACE FOR ANSWERS

[Turn over

SPACE FOR ANSWERS

ADDITIONAL GRAPH PAPER FOR QUESTION 11(*c*)

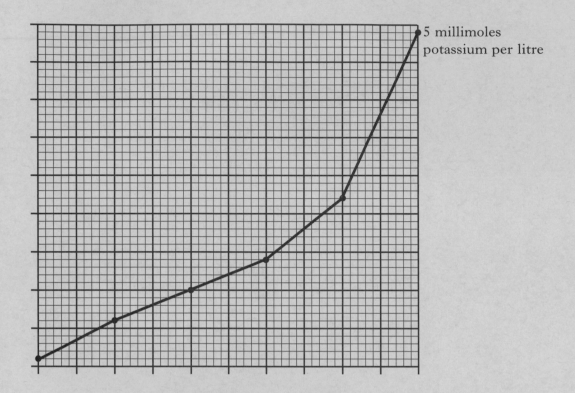

5 millimoles
potassium per litre

[BLANK PAGE]

[BLANK PAGE]